THE HIPPOS

THE
HIPPOS

Natural History and Conservation

S K Eltringham

Illustrated by Priscilla Barrett

T & A D
POYSER
NATURAL
HISTORY

Text © 1999 by T & A D Poyser Ltd
Illustrations © Priscilla Barrett

Academic Press
24–28 Oval Road, London NW1 7DX, UK
http://www.hbuk.co.uk/ap/

Academic Press
A Harcourt Science and Technology Company
525 B Street, Suite 1900, San Diego, California 92101-4495, USA
http://www.apnet.com

ISBN 0-85661-131-X

A catalogue for this book is available from the British Library

Typeset in 10pt Bembo by Phoenix Photosetting, Chatham, Kent
Printed in Great Britain by University Press, Cambridge

99 00 01 02 03 04 CU 9 8 7 6 5 4 3 2 1

CONTENTS

The plate section can be found between pages 88 and 89.

LIST OF PLATES

1. Hippos and crocodiles feeding on the carcase of a hippo. An unusual case of cannibalism. (Photo N. Dunnington-Jefferson)

2. Hippos tearing at the flesh of a hippo killed in a fight. (Photo N. Dunnington-Jefferson)

3. A female hippo and her well-grown calf.

4. A school of hippos in a permanent river.

5. A group of mostly male hippos in a temporary wallow.

6. The result of over-stocking with hippos. Bushes have encroached onto what was once grassland.

7. A badly eroded lake shore as a result of hippo grazing.

8. Lion Bay in Queen Elizabeth National Park, Uganda, after removal of hippos. Previously it was in the same condition as that shown in Plate 7.

PREFACE

During the time when I was living and working in the Queen Elizabeth National Park in Uganda, hippos were an inescapable part of my life. There were some 12 000 or more in the park including some large schools close to my house. I became used to falling asleep to their accompanying grunts and snorts and sometimes to be awoken by the sound of munching from one that had wandered into my garden. They acted as lawn mowers, which was just as well as there was no chance of growing anything other than grass and any attempt to cultivate, which was, in any case, not allowed in the park, would have suffered from crop-raiding. When I first arrived, a detailed study of the species had just been completed, based on specimens removed in the culling exercise described in this book. The purpose of the culling was to restore the vegetation, which had suffered from overgrazing when hippo numbers had climbed to around 21 000. Obviously the Park's authorities wanted to know if the experiment had worked and consequently, there was plenty of work for our botanists to do. My interest as a zoologist, was in the effects that the reduction in hippo numbers may have had on the other species of large mammals. This involved me in repeated game counts as well as in aerial surveys of the hippos themselves to check on their recovery from such a massive perturbation. I soon fell under the sway of these irascible and dangerous, yet loveable, creatures and I regretted that because of other duties I was unable to find more time to devote to them but I always hoped that one day I would be able to investigate their biology in more detail.

I did not abandon them on my return to Britain for I became Chairman of the Hippo Specialist Group, one of the many set up by the International Union for the Conservation of Nature and Natural Resources (IUCN), or the World Conservation Union, as it is now called. The purpose of the groups is to keep an eye on the welfare of the species concerned and to advise IUCN on their conservation. This involves the production of an Action Plan by each group and in doing this for hippos, I was obliged to investigate their numbers and distribution throughout their range, which means sub-Saharan Africa. The results are included in the following pages but having got so far with hippos, I thought that I should attempt to bring together as much as possible of what is known about the species. I was surprised to find that this had not been done before but although hippos are considered in more general works there are some excellent picture books available, I could not find any text that considered only hippos. Consequently I was pleased to have the opportunity of filling that gap. I hope this book will be read by the general public as well as by biologists as I have tried to avoid too much technical detail. It is not intended to be a comprehensive review of the literature and the reference list includes only part of the many papers and books I have read in preparing this text.

Much of the important research on the reproduction and ecology of hippos was carried out over thirty years ago but most of it has never been superseded and remains relevant today. The probable reason is that there are fewer specimens available for dissection from culling schemes nowadays. On the other hand, the study of hippo behaviour is more recent and some fundamental discoveries have been made. There is still much to learn about hippos and progress may be expected in the analysis of their DNA in order to disentangle their confused ancestry. One can also expect advances from studies of the ways in which they communicate with each other.

I have received help from several quarters in writing this book. It would not have been possible to carry out the continent-wide survey of the distribution and numbers of hippos but for the enthusiastic assistance of many observers throughout Africa. Their names occur in the section on the various countries in Chapters 10 to 12. I was ably supported by William Oliver, who gave practical advice in organising the questionnaires and, with Alastair Macdonald, critically reviewed drafts of the Action Plan on which the relevant chapters in the present book are based. Some of the other chapters were read by colleagues and I particularly thank Eleanor Weston for criticising the chapter on fossil hippos and Michael Lock for commenting on the sections on the ecology of hippos. I am grateful to my wife for reading the whole of the book in proof and for picking up numerous typos and errors. My thanks are also due to Miss Nicky Dunnington-Jefferson for telling me about the case of cannibalism in Malawi and for allowing me to reproduce her photographs of the scene. I am indebted to the following societies and publishers for permission to reproduce material from their books and journals: the British Ecological Society (Table 7.3, Figs 2.2, 2.3, 5.1, 5.5, 7.1, 7.5, 7.6, 7.7), Blackwell Science Limited (Tables 2.1, 4.1, 6.1, 7.1, 7.2, Figs 2.7, 5.6, 5.8, 7.2), K. Hentschel of Brunswick University (Table 6.3), Cambridge University Press (Table 8.1), Centro di Studio per la Faunistica & Ecologia Tropicali (Fig. 11.1), East African Natural History Society (Fig 11.2), Harvard University Press (Figs 2.5, 2.6), IUCN (Fig 10.2), Mammalia (Fig. 4.1), Munksgaard International Publishers Ltd. (Table 2.9), National Parks Board of South Africa (Fig. 5.2), Quartarpaleontology (Fig. 3.1), the Physiological Society (Table 2.2), the Southern African Wildlife and Environment Society (Fig. 2.9), K. Susuki (Figs. 7.8, 7.9), TRAFFIC International (Fig. 9.1), Wiley-Liss Inc., a subsidiary of John Wiley & Sons Inc. (Fig. 2.10), A. Ziemsen Verlag (Fig. 2.4), The Zoological Society of London (Tables 2.4, 2.6, 2.7, 2.8, 2.10, Figs 2.11, 5.3, 5.4, 5.7, 10.1) and the Zoological Society of Southern Africa (Table 6.2, Fig. 2.12). The authors are also thanked for permitting me to use their data. They are named in the captions to tables and figures. The staff of Academic Press have been most helpful and I thank in particular Andrew Richford and Manjula Goonawardena.

S.K.E.
Cambridge

Note: After the bulk of the book was written, the name of Zaire was changed to the Democratic Republic of the Congo. I have not thought it necessary to alter all references to Zaire in the text.

CHAPTER 1

INTRODUCTION

The two living hippopotamus species

The hippo is one of the best known animals, recognisable at a glance, and is a great favourite with the general public, not least with children, many of whom possess a cuddly hippo toy. It is also a source of adult humour and hardly a week goes by without the appearance of a hippo cartoon in one or other of the magazines or newspapers. Thanks to Flanders and Swann, everyone knows that hippos enjoy wallowing in mud. Many zoos exhibit hippos and most people in the world have seen one in the flesh. It has featured in human affairs since at least the time of the Pharaohs, when it was venerated as a god, and it has been portrayed in art down the ages. Early records of human interest in the animal can be seen in the rock paintings and engravings of hippos, amongst other large mammals, in the mountains of the central Sahara. One near Djanet in the Tassili n'Ajjer Mountains, dating from 4000–5000 years ago, appears to represent a hippo hunt with three hippos and three canoes. Apart from anything else, this find shows that the Sahara was a well watered region in the not so distant past.

Despite the hippo's popularity, not much is known about its biology and what little that has been found out often clashes with its jolly, roly-poly public image. As one who has been chased by a hippo on a number of occasions, both on land and water, I can attest to its sometimes short temper. Because of its aquatic and nocturnal habits it is not the easiest of animals to study, yet there has been a considerable body of research published on the hippo, about which the general public is mostly unaware, and the time seems ripe to gather this information into an overall survey of the biology of the species. Good general accounts of hippo biology are given by Kingdon (1979), Macdonald (1984) and Klingel (1988).

Hippos are traditionally thought to be distant relatives of the pigs and peccaries, with which they are grouped within the sub-order Suiformes. Recent research has thrown some doubt on this relationship. There are only two species of hippos alive today but not so long ago there were several more. The two living species are the large common hippo, sometimes called the river hippopotamus, *Hippopotamus amphibius*, and the much smaller pygmy hippo, *Hexaprotodon liberiensis*, which is also a familiar zoo exhibit. They are two quite different animals, so unalike, in fact, that they have been assigned to different genera.

Several other species of hippos, generally, but not universally, considered as belonging to the genus *Hippopotamus*, lived in the Pleistocene, i.e. around a million years ago, with two species in Britain and perhaps as many as eight species in Africa. Some survived until quite recently on Madagascar. Two of these were pygmy forms but the third, *H. laloumena*, was a full-sized creature. The small animals were *H. lemerlei* and *H. madagascariensis*. The former lived to about 1000 years ago and was probably killed off as a result of human settlement on this previously uninhabited island but the other two species have not been dated although they are believed to be of Holocene age, i.e. to have existed within the last few thousand years. Their small size does not mean that they were necessarily related to the living pygmy hippo for island dwarfing is a well known phenomenon in mammals. Details of the structure and origin of these and other extinct hippos will be given later.

THE COMMON HIPPO (*HIPPOPOTAMUS AMPHIBIUS*)

The common hippo is now confined to mainland Africa south of the Sahara although its distribution is much more restricted than it was only a few decades ago. The distribution and numbers of hippos are considered in Chapters 10–12. Five subspecies of *Hippopotamus* have been recognised based on skull shape and proportions (Grubb, 1993) but it is not certain that these represent anything more than a classification of museum specimens. There may well be geographical variations in skull features but they are not obvious in the field as most of the alleged differences are relative. Hippos are split up into a number of separate populations which probably differ genetically but whether or not they constitute separate subspecies is largely a matter of opinion. It is important for conservation reasons, however, to preserve these gene pools, whatever they are called. Although no field biologist recognises hippo subspecies nowadays, their names and distributions are given below for the sake of completeness. The availability of modern DNA fingerprinting techniques could help to disentangle the relationships but, as far as I know, the method has not yet been used with hippos.

H. a. amphibius: The nominate race is said to have occurred in Egypt, where it is now extinct, and as far south as Tanzania and Mozambique.

H. a. tschadensis: This race is found in west Africa including, as the name suggests, Chad.

H. a. kiboko: *Kiboko* is the Swahili name for the hippo. This subspecies is from Kenya and Somalia.

H. a. constrictus: The range of this race is Angola, southern Zaire and Namibia. The name *constrictus* refers to the deep preorbital constriction.

H. a. capensis: This is the southern African hippo.

The name Hippopotamus literally means river horse from the Greek *hippos* – horse – and *potamus* – river – but there is very little that is horse-like about a hippo. The allusion to horses presumably originated from the appearance of the eyes, ears and nostrils, which may protrude above the water when the animal is submerged. With a little imagination, one can turn these structures into a horse's head. The second part of the name, *potamus*, is certainly appropriate as rivers are favourite habitats of the hippo. The animal is truly amphibious and is well called *Hippopotamus amphibius*. It spends the day in the water but it does not eat aquatic vegetation to any extent. It is, in fact, a nocturnal grazer, which emerges at dusk to feed up to several kilometres from its day-time resting place. It follows regular paths to do this and, over time, distinct trails are formed. Sometimes they are so eroded that virtual trenches result. These can be significant in swamps as they help to permit the flow of water (McCarthy *et al.*, 1998).

The common hippo is unmistakable with a barrel-shaped, almost hairless body weighing in at anything up to 3 tonnes although half that weight would be a more average figure. It has stout, but disproportionately short, legs. Nevertheless, the hippo can run at fast speed on land and it would be futile to try to outpace one that was in an aggressive mood. The feet are partially webbed but, funnily enough, the hippo is not a good swimmer. It could be said that it doesn't swim at all as it is rarely found out of its depth. It cannot float and those animals seen on the surface are really standing in the shallows or, more likely, lying down. When they are in deep water, they progress by a series of porpoise-like leaps off the bottom in a curiously graceful manner reminiscent of an overweight ballet dancer filmed in slow motion. Alternatively, they may move with a series of high, prancing steps. It is, of course, the support given by the water that allows the hippo to progress in such an elegant fashion. Despite its plump appearance, the hippo has only a thin layer of subcutaneous fat.

The hippo is not streamlined for rapid progress through the water. Professor Frank Fish of West Chester University, Pennsylvania, has calculated the Fineness Ratio (FR) as a measure of streamlining and water flow patterns for a number of aquatic mammals. The FR is the body length divided by the girth and has high values for fast swimming animals. It can be over 10 for some dolphins but the optimal value for a body of maximum volume and minimum hydrodynamic drag is 4.5 (F. Fish, *in litt.*). The FR for hippos can be calculated from measurements of some Uganda specimens kindly supplied by Dr R.M. Laws (*in litt.*). Values can be as low as 2.5 suggesting that the body dimensions have not evolved to maximise speed through the water.

Underwater activity is rarely seen, as the hippo is usually in muddy water, but occasionally, such as at the Mzima Springs in Tsavo National Park, Kenya, the water is crystal clear and the gambolling of submerged hippos may be watched from an under-water observation chamber. The maximum time that a hippo can remain under water is not certain as, for obvious reasons, no-one has carried out the ultimate experiment of seeing how long it takes to drown one. The period that a hippo normally spends submerged before surfacing to breathe has been recorded on several occasions, however, and the figure generally given is from four to six minutes. No doubt it could stay submerged for longer if it were forced to do so. The calves breathe more frequently, perhaps surfacing every two or three minutes. They often climb onto their mothers' backs if the water is too deep for them to stand up in it.

A reflex action ensures that the nostrils and ears are closed by muscular valves as soon as they come in contact with water. Air in the lungs is expelled in an explosive burst on surfacing, sending a miniature water-spout into the air. At the same time the ears are waggled furiously to clear them of water. If danger threatens, however, a hippo can exhale so quietly that it can hardly be heard from the bank.

Hippos will often leave the water during the day to bask in the sun but they never go far away so that they can easily slip back into the water if they become over-heated. The skin is very sensitive and will crack if exposed to direct sunlight for long. The hippo is physiologically dependent on water but liquid water is not essential and it can manage quite well provided the skin is kept moist – hence its predilection for mud. Hippos can be found in all types of water, from rivers and lakes to muddy wallows. They can even occur off the sea shore.

The hippo does not spend all its time in water but comes ashore at night to graze on grasslands which may be several kilometres inland. Groups of hippos seen in the water by day are usually referred to as "schools". This vague term reflects the general ignorance of hippo behaviour but it is now becoming clear that hippos in the water are territorial and that schools consist of groups of females, under the control of a territorial male, and bachelor groups. It is unlikely that the females in these groups represent social associations or that they consistently return to the same territory each morning, at least in the long term. The bulls are not territorial away from water and all animals feed on their own rather than in groups, although calves accompany their mothers until they are well grown. Other males may be found within the territory but they behave submissively towards the territorial bull, which is easily recognised from its confident and aggressive behaviour.

Sometimes a hopeful youngster will challenge the territory holder and if the animals are well matched, a titanic battle ensues during which extensive injuries may be inflicted by the enormous canine teeth, often leading to the death of the defeated animal. Confrontations between territorial bulls are highly ritualised and it is only when the territory is at stake that serious fighting occurs. Most old bulls carry extensive scars from past conflicts but despite the severity of the wounds, which may penetrate the peritoneum, sepsis rarely occurs and healing is usually swift and complete. It is possible that the red secretion from skin glands acts as an antibiotic. Nevertheless, deaths can result, probably from loss of blood, if two equally matched bulls are involved. The huge canine teeth are the principal weapons of hippos and they are prominently displayed during the incorrectly termed "yawning", which is really a warning signal. For the most part, however, the relationships between the territorial bulls and the other males, particularly the young ones, are extremely friendly (Klingel, 1991).

An amusing aspect of hippo behaviour is "muck-spreading". During defaecation, the hippo wags its short tail vigorously, which has the effect of spreading the dung over a wide area. This behaviour occurs both in the water and on land, where it is usually a bush that is liberally sprayed. The function is not clear. It does not appear to involve territorial marking as hippos are not territorial on land, but it may help to guide the animals back to water after their feeding sessions. The possibility that dunging has a social significance is considered in Chapter 4.

Courtship is a rough affair, with the male forcing its attentions on the female. Hippos mate in the water and usually give birth there as well. Suckling also takes place under water. The mother goes off on her own before the birth and remains away from her group for about ten days after the calf is born (Klingel, 1991).

The hippo has a deep resonant call that would be menacing were it not preceded by a near falsetto squeal. Bulls, when fighting, emit furious, loud bellows. A recent discovery about hippos is that they vocalise under water as well as on land (Barklow, 1995). Some of the sounds are similar to those made on land except that they may be much louder. Others are made only under water and include grunts, clicks and whines. These may serve for communication but there is no evidence that hippos use clicks to echolocate under water as dolphins do. Some calls are amphibious in that they are transmitted simultaneously through both air and water. It is not yet clear what function these sounds have.

Hippos are difficult to count because of their habit of sinking and remaining submerged when they are approached. When they do surface to breathe it is often only the tops of their heads that are visible. There have been several counts of hippos in various parts of their range (Sidney, 1965) but no attempt was made to estimate the total throughout Africa until my effort in 1989 (Eltringham, 1993a). This was carried out largely through correspondence but was based on real counts in many places. Inevitably, the quality of the data collected varied considerably but the total is probably of the right order of magnitude. Altogether 157 000 hippos were enumerated. This figure has been slightly modified upwards to nearly 174 000 in a re-analysis made here in Chapters 10–12. There are places where hippos still occur in abundance but there are others which have witnessed a marked reduction in both the numbers and geographical range of hippos. This is particularly true of West Africa, where only a few remain in scattered, remnant populations. Largest numbers occur in East and Central Africa, with Zambia, eastern Zaire and Malawi holding particularly high numbers.

Hippos are long-lived animals although by 40, they are decidedly geriatric and not many live beyond that age in the wild.

THE PYGMY HIPPO (*HEXAPROTODON LIBERIENSIS*)

Unlike the common hippo, which has been a familiar animal since classical times, the pygmy hippo was not known until well into the 19th Century. It was first described scientifically in 1844 by a Dr Morton from skulls received when he was resident in Monrovia. He first named it *Hippopotamus minor* but altered the name to *Choeropsis liberiensis* in 1849 when its distinctive features were recognised. More recently the scientific name was changed to the present *Hexaprotodon* after Coryndon (1977) showed that the pygmy hippo is related to extinct forms of *Hexaprotodon* and should, therefore, be included with them. The name *Hexaprotodon* means "six front teeth" and refers to the possession of three pairs of incisors in some fossil forms. The living *Hexaprotodon*, however, has lost two pairs and, technically, is diprotodont. Incidentally, the old name, *Choeropsis*, comes from the Greek *Khoiros*, meaning a pig or hog, and *opsis* meaning "belonging to".

The pygmy hippo, weighing between 180 and 275 kg, is much smaller than the common hippo. It is nevertheless quite a large animal and is treated with respect by hunters on account of its aggressiveness at close quarters. The species is mainly found in Liberia, as might be guessed from the second part of its scientific name, but small populations may still occur in the neighbouring states of Guinea, Ivory Coast and Sierra Leone. It was always confined to west Africa and it is likely that its distribution has not changed much since prehistoric times. A separate subspecies (*H. l. heslopi*) once existed in Nigeria but there have been no recent sightings and it may well be extinct. The report of a further population in Guinea Bissau (Cristino & Melo, 1958) was based on a shot specimen which was almost certainly a young common hippo.

The existence of the Nigerian subspecies means that the full scientific name of the Liberian animals is *Hexaprotodon liberiensis liberiensis*. Within Liberia, the pygmy hippo is quite widely distributed although it is nowhere abundant. It is much more difficult to count than the large hippo and the warfare in the country at the time of my census precluded any direct counts of the animals. However, the general opinion, based on no hard evidence as far as I can see, is that some 2000 remain. Roth *et al.* (1996) give a much more optimistic estimate of many thousands based on extrapolation from sample counts.

The range of the pygmy hippo does not overlap with that of the common hippo except, perhaps, in the upper stretches of some rivers in the Ivory Coast where forest has given way to agriculture. A few common hippos have settled there and it is possible that the two species may occasionally meet (Roth *et al.*, 1996).

It is a very different animal, ecologically, from the common hippo, being an inhabitant of forests rather than grasslands. It is more like a pig than a hippo in its ecology but Robinson (1970) believes that it shows a closer ecological similarity to tapirs. Although it is never found far from water, it is much less aquatic than its larger relative. In its general anatomy, the pygmy hippo is similar to the common hippo but there are some obvious differences. Whereas the back of the larger hippo is more or less parallel with the ground, the pygmy hippo's body slopes markedly forward, possibly as an adaptation to facilitate its passage through thick vegetation. Its limbs and neck are proportionately longer and its head smaller with less protuberant eyes and nostrils. It is also more delicate, with a relatively smaller skull and a more slender skeleton. There are distinct anatomical adaptations for terrestrial life in the limbs, with their limited lateral rotation, and in the narrower feet, with shorter lateral digits. The toes, which are even less webbed, spread out sideways to a greater extent than in the large hippo. This is probably a terrestrial adaptation that assists walking on the soft forest floor. The pygmy hippo, nevertheless, has such aquatic adaptations as strong muscular valves to the ears and nostrils and a water-dependent skin physiology as complex as that of the larger species.

The pygmy hippo feeds mainly at night on fruits and other forest products and spends most of the day hidden in wallows or swamps. It is believed to lie up sometimes in subterranean chambers within river banks although it is not known to construct such burrows (Robinson, 1981). It is not very social for it rarely occurs in groups of more than two or three. Its nocturnal, solitary habits, coupled with the dense nature of its habitat, make pygmy hippos very difficult to see and it is hardly surprising that not very much is known about their habits. The most recent study of wild pygmy hippos was that carried out in the early 1980s in the Ivory Coast by Hentschel (1990).

The life span in captivity varies from 42 to 55 years but it is unlikely that many pygmy hippos reach such ages in the wild. According to the compilers of the pygmy hippo stud book (Tobler, 1988), the first specimen to be seen in Europe was an infant caught in Sierra Leone soon after birth and shipped to Liverpool in 1873 by one John Price of the British Colonial Service. It was destined for Dublin Zoo but it seems that it suffered from too much mollycoddling in being over-protected from cold and it did not survive for long. It would have done better if it had been allowed to become acclimatised. A more successful introduction of pygmy hippos into Europe was made in 1911 by Major Hans Schomburg, who caught several specimens in Liberia using pit traps and shipped them first to Germany and then to the Bronx Zoo in New York, where they appear to have done very well and bred successfully. It is now a familiar zoo animal and, ironically, its future in captivity seems more secure than it is in the wild.

In a popular article, Philip Robinson recounts some interesting folk tales about the pygmy hippo. One is that the hippo carries a diamond in its mouth at night in order to light its way through the forest. It hides the diamond during the day in a place where it cannot be found but any hunter lucky enough to kill a hippo at night can keep the diamond. Another story is that the baby hippo is not suckled in the normal way but is fed on the foamy secretion on the mother's body, which is licked off. It is easy to see how this belief could have originated for the teats are not conspicuous and the skin secretions often become worked up into a lather after exertion.

CHAPTER 2

ANATOMY AND PHYSIOLOGY

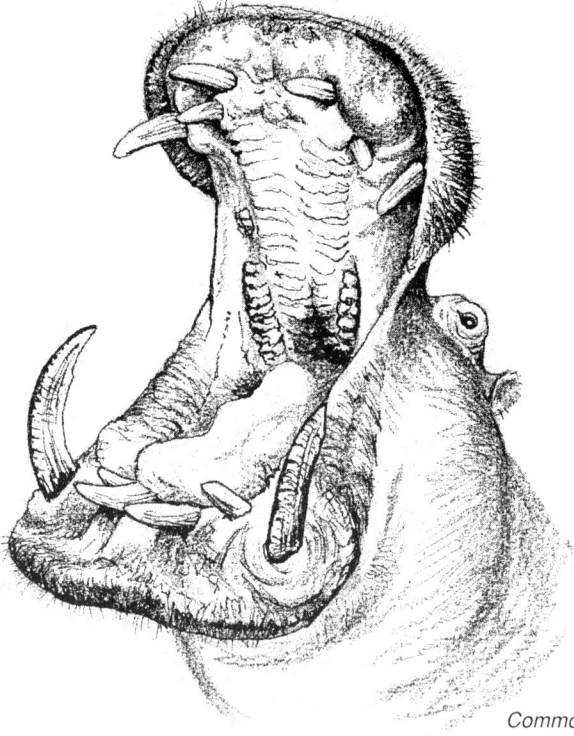

Common hippo 'yawning'

M ost of the following account applies to the large hippo *Hippopotamus amphibius* but differences in the pygmy hippo from the condition described will be pointed out where appropriate.

ANATOMY

Skeleton

The common hippo has a skeleton that is known as graviportal, that is it is "designed" to support a great weight (Fig. 2.1). The bones are massive particularly the vertebrae of the spine, which form a rigid girder from which the soft tissues of the gut and internal organs hang. The 13 pairs of stout ribs constitute a barrel-like cage, which helps to give the hippo its rotund appearance. The scapula, or shoulder blade, is orientated

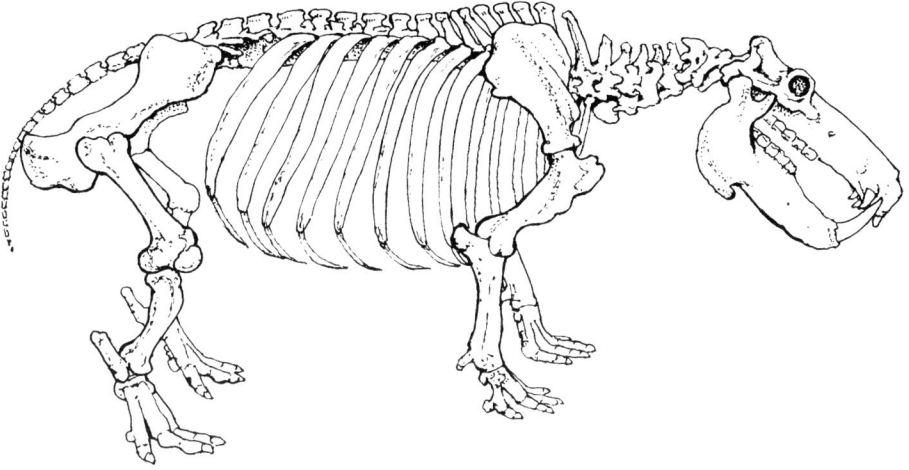

Figure 2.1 Skeleton of common hippopotamus.

vertically and, with the leg bones, forms a vertical column, another graviportal feature. The rear legs are not involved in a similar column as the pelvic girdle is not upright but lies at a 45-degree angle. Although undoubtedly robust, the limb bones are, in fact, rather less heavy than might be expected in such a large animal and the reason may be that much of the weight of a hippo is taken by the water. Like all artiodactyls, hippos have an even number of digits but they differ in that there are four well developed toes on each foot, rather than two as in other members of the order. Each toe is finished with a broad nail rather than a hoof. Also, like other artiodactyls, the hippo walks on its toes, i.e. it is digitigrade and consequently, there is a distinct hock to the hind leg. The gait is not as exaggeratedly digitigrade as in, say, an antelope and some authors describe it as semi-digitigrade.

The post-cranial skeleton of the pygmy hippo is very similar but is more "gracile", a word which, by comparison with the big hippo, might be interpreted as dainty. It is not, however, merely a scaled down version of the larger species. There are differences in the skull, which will be considered below, as well as in the skeleton but those of the latter are of a technical nature outside the scope of this book. Some of the changes are allometric, e.g. the organs do not always show the same degree of dwarfing in extinct hippos that evolved a smaller size on islands.

Weight

The common hippo is a huge animal and qualifies as a megaherbivore according to the criterion of Owen-Smith (1988), i.e. over 1000 kg. The hippo easily exceeds this limit, with a weight of up to 3200 kg reported for an exceptionally large male. That

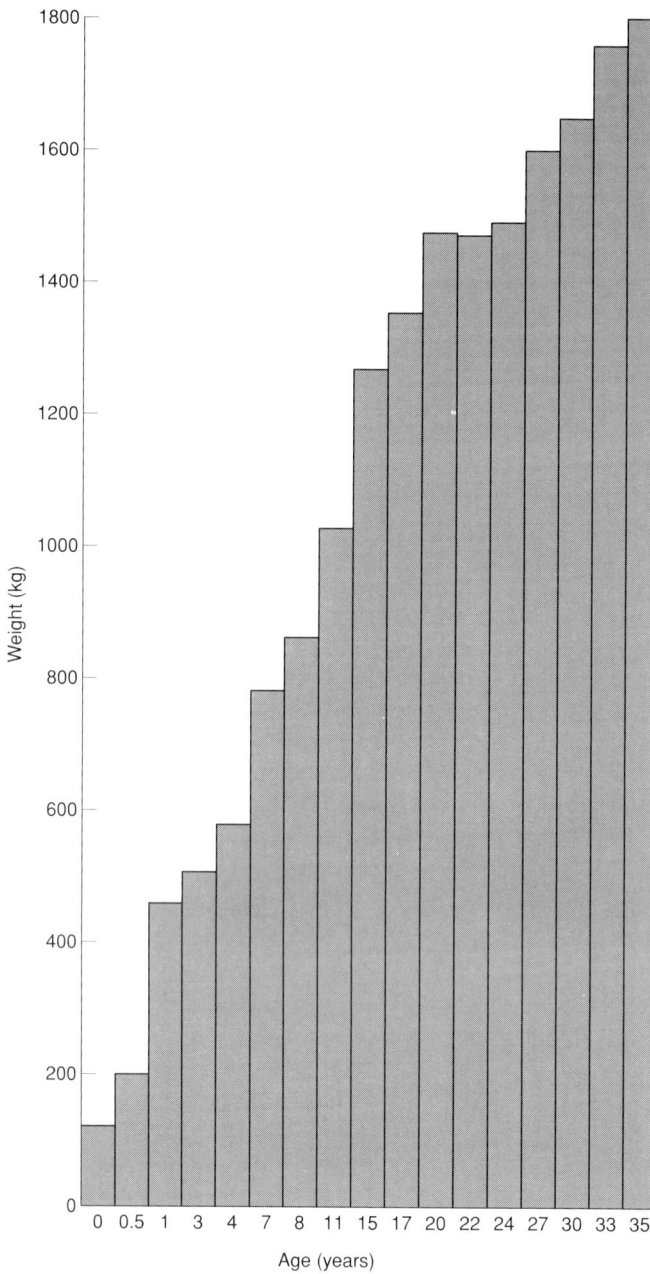

Figure 2.2 Weight in relation to age of 189 male hippos shot in the Luangwa Valley, Zambia, in 1970. This is equivalent to a growth curve, which suggests that male hippos continue to grow throughout life for there is no suggestion of the curve levelling off. After Marshall & Sayer (1976).

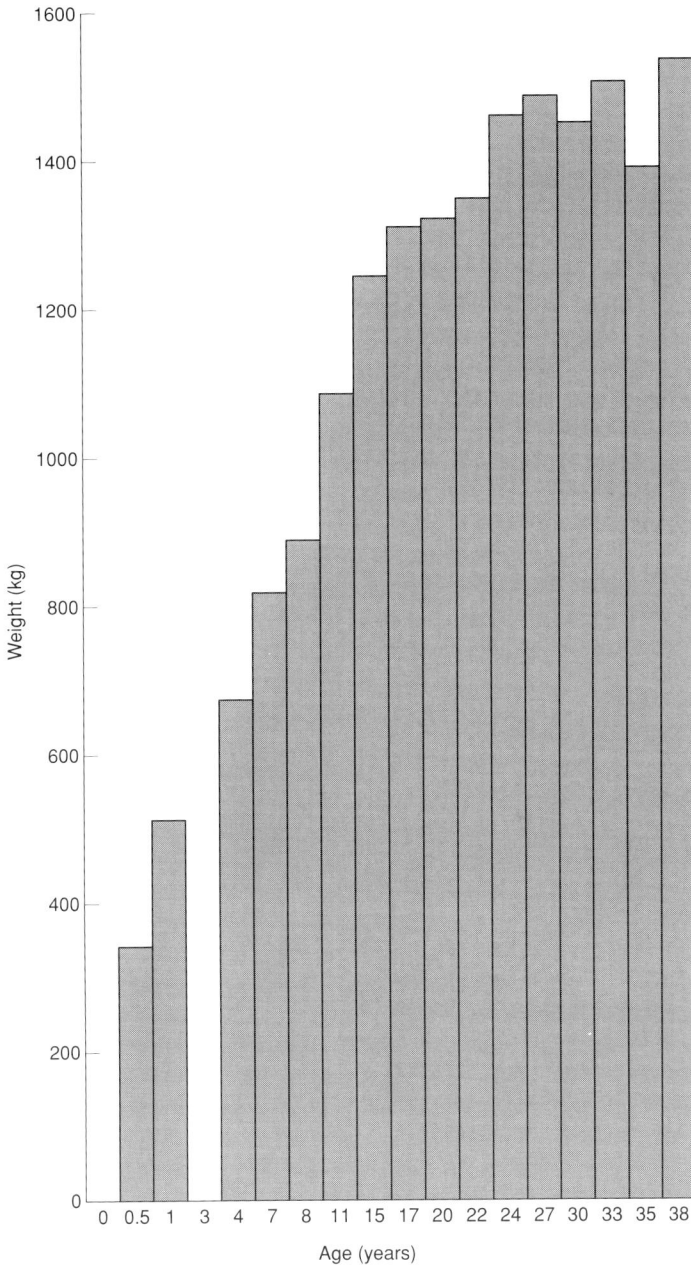

Figure 2.3 Weight in relation to age of 186 female hippos shot in the Luangwa Valley, Zambia, in 1970. The female hippo appears to approach asymptote at 24 years of age and to stop growing, unlike the male, which continues to grow. After Marshall & Sayer (1976).

seems rather high but it is difficult to trace an authoritative weight for a hippo and most figures given in field guides and the like merely state a figure without any supporting information. It is not easy to weigh a wild hippo, short of killing it and chopping it up into manageable portions, and few zoos have weighbridges of sufficient capacity to weigh living animals. Where culling has been carried out and actual weights recorded from dismembered corpses, the results have been less spectacular than some of the previously published figures, e.g. Laws (1963), who weighed a large sample in Uganda, found the average weight to be 1536 kg for males (max. 2065 kg) and 1386 kg for females (max. 1716 kg). Ledger (1968), who weighed four specimens of each sex, reported similar average weights for East African hippos – 1490 kg for males (range 1179–1714 kg) and 1277 kg for females (1185–1401 kg). Pinaar *et al.* (1966) found similar weights in the Kruger National Park, South Africa, with means of 1490 and 1321 kg for males and females respectively and corresponding maxima of 1999 and 1674 kg. Even if these values fall short of some of the exaggerated estimates, hippos are heavyweights by any standard.

Marshall & Sayer (1976) weighed several hundred hippos in the Luangwa Valley, Zambia, in 1970 and 1971 and present some tables showing average weights at various ages. Individual weights are not given but the mean values were below 1800 kg for adult males and under 1500 kg for females, although a few females exceeded that figure. The values, which are plotted in Figs 2.2 and 2.3, can be treated as growth rates. The female growth seems to reach asymptote around 24 years of age but there is little evidence that the males' growth has begun to level off, suggesting that males continue to grow throughout life. The sample of old hippos is small, however, and further measurements are required before this tentative suggestion can be accepted as true.

The pygmy hippo is a much lighter animal weighing around a sixth of the weight of the larger species. Kingdon (1997) gives the weight as ranging from 180 to 275 kg without quoting original sources. Lang (1975) quotes some weights for animals at Basel Zoo. Four females there, aged 6, 7, 14 and 29 years, weighed 213, 243, 267 and 247 kg respectively. The weight of another cow was 179 kg at death and that of a bull 192 kg although the weight in life was said to be 208 kg. An 18-year-old bull weighed 273 kg, which is probably a better estimate for the weight of an adult male and suggests that, as in the large species, the male is slightly heavier than the female.

Lang shows the average growth rate of three female pygmy hippos and one bull over the first five years of life (Fig. 2.4). This approximates to a straight line although there is a suggestion of a sigmoid curve, which would be expected in a graph of this nature.

Size

Some measurements were made by Dr R.M. Laws of the lengths of hippos shot on control in Uganda during the 1960s. The sample contained 20 of each sex. The average length along the back was 269.5 cm (about 8ft 10in) for males and 271.75 cm (also about 8ft 10in) for females. The measurements ranged from 119 cm to 302 cm in males and from 183 cm to 302 cm in females.

The girth dimension was also measured by Laws. Because of the difficulty of passing a tape measure under the carcase of a hippo, the measurements were taken over the back from behind the forelimbs and multiplied by two on the assumption that this

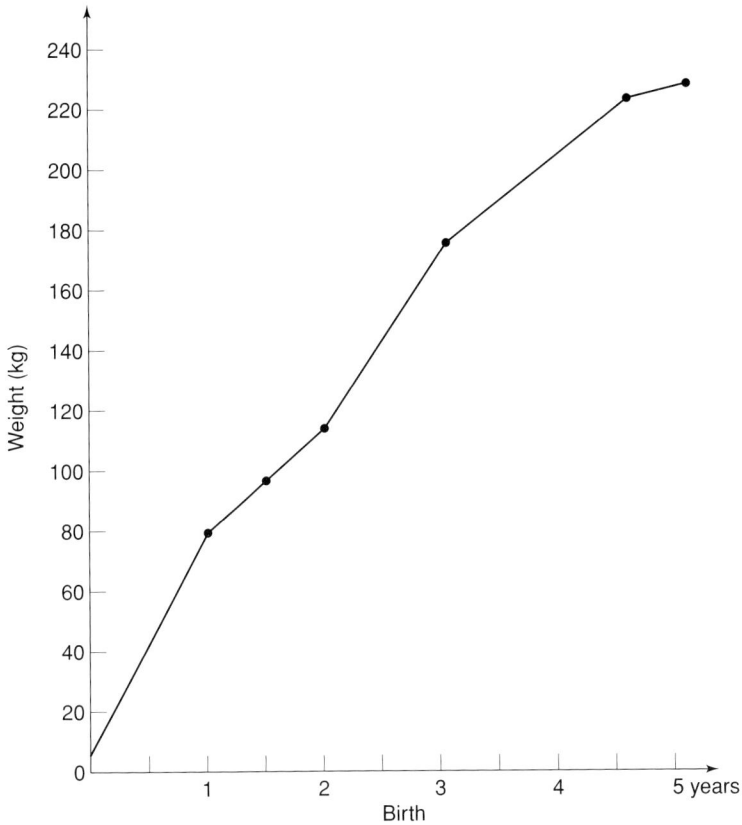

Figure 2.4 Average growth rate of three female pygmy hippos and one male born in Basel Zoo, Switzerland. The asymptote is not reached suggesting that growth is not completed by the age of five years. From Lang (1975).

measured the half girth. In any case full girth measurements of dead animals might well have been distorted by bloat. It is interesting to note that the girth was found to be very similar to the length of a hippo. The average figures were 269.2 cm for males and 268.5 cm for females with respective ranges of 157–315 cm and 152–335 cm. There is, therefore, very little differences between the sexes in linear measurements but the males are about 10% heavier.

The body length of the pygmy hippo according to Kingdon (1997) is 150 to 177 cm (4ft 11in to 5ft 10in) but the upper limit seems rather high compared with measurements made by Lang (1975) of captive specimens. He gives the lengths of three adult females as 142, 150 and 150 cm, respectively, and that of a male as 157 cm. As with weight, there is probably a tendency for the male to be larger although the height of the male at the hindquarters was only 81 cm (2ft 8in) whereas that of one of the females was 83 cm. The sample sizes for all the pygmy hippo measurements are obviously very small and can give no more than an approximation of the average values.

Teeth

The teeth of hippos serve two quite different functions; one is feeding and the other is fighting. Feeding is carried out with the back or cheek teeth and fighting with the front teeth. The dentition of the earliest fossils was complete with a set of 44 teeth, the maximum to be found in terrestrial mammals, but the number of permanent teeth is reduced to 36 in the living *H. amphibius* and to 34 in the pygmy hippo.

Canines

The front teeth of the common hippo consist of four massive, tusk-like canines and eight shorter, peg-like incisors, four in each jaw. Both types have open roots and so continue to grow throughout life. The upper and lower canines abut and are kept sharp by grinding together, rather like the incisors of a rodent. If one canine is lost, its partner continues to grow in a spiral, and can seriously interfere with feeding. One captive hippo that I saw had lost an upper canine and the lower one had penetrated the upper lip so that it was visible when the mouth was closed. The canines in the upper jaw are much shorter than those in the lower although they are equally sturdy and sharp.

The canines, which are conspicuously ridged lengthways, may, in the lower jaw, project for about 30 cm (1 ft) from the gum but there is a longer root, some 40 cm in length, bringing the total to 70 cm (about 2ft 4in). Such large teeth are approaching elephant tusks in size and have a considerable trophy value as well as being suitable for carving, although their triangular cross-section restricts the designs that can be fashioned. Not all teeth are of such large dimensions, of course, but even small specimens can be incorporated into attractive *objets d'art*.

Incisors

The incisors are not so big as the canines but their size is still impressive. Those in the lower jaw are longer and project forward for up to 17 cm (6–7 in) from the gum. The outer incisors of the common hippo are larger than the inner pair. The incisors in the upper jaw are round in cross-section whereas the lower incisors are triangular. The upper incisors also differ in pointing downwards as well as being shorter. Both types have blunt tips.

The pygmy hippo differs in having only one pair of incisors in the lower jaw rather than two. There is also a species difference in the insertion of the incisors in the upper jaw (maxilla). In the large hippo, the incisors lie one behind the other and contact the lower incisors in a scissor-like motion. By contrast the upper incisors of the pygmy hippo lie more or less side by side and meet the lower incisors tip to tip.

The canines, and possibly the incisors, are used for fighting. None of the front teeth plays any part in feeding although the incisors are said by Laws (1968a) to be used for digging, including the mining of minerals from salt licks. There is a marked sexual dimorphism in the canines, which are much larger in the male.

Grinding (molariform) teeth

The teeth that are used in feeding are the back teeth, six in each half jaw, comprising three premolars and three molars in the permanent set. The pattern of cusps on the molar teeth has changed little over evolutionary time and is unique to the family

Hippopotamidae. The molars are roughly square with a cusp in each quarter. These wear down during life into characteristic shapes which identify the species (Fig. 2.5). The arrangement of the teeth also differs between the species, with the length of the premolar row being longer, relative to the molars, in the pygmy hippo (Fig. 2.6).

Milk teeth

The milk teeth of the large hippo consist of four premolars in each half jaw but the first one is not replaced in the adult set of teeth although it can be retained for some time. The adult may, therefore, appear to have four permanent premolars but there are really only three. When it is retained, the milk premolar is small, almost vestigial, and is separated from the other premolars by a wide gap or diastema.

The milk incisors and canines are usually erupted at birth but they are lost within the first few months of life. There are three milk incisors in each half jaw but only the first and second are replaced by permanent teeth. The pygmy hippo has a very similar dentition but only one pair of the lower milk incisors is replaced in the adult.

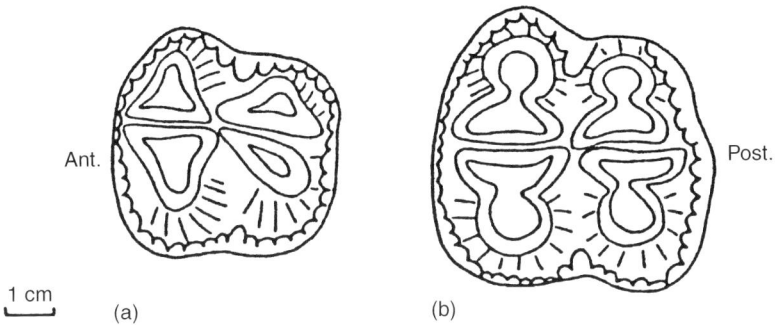

Figure 2.5 Partially worn third upper molar of the pygmy hippopotamus (a) and the common hippopotamus (b). From Coryndon (1978).

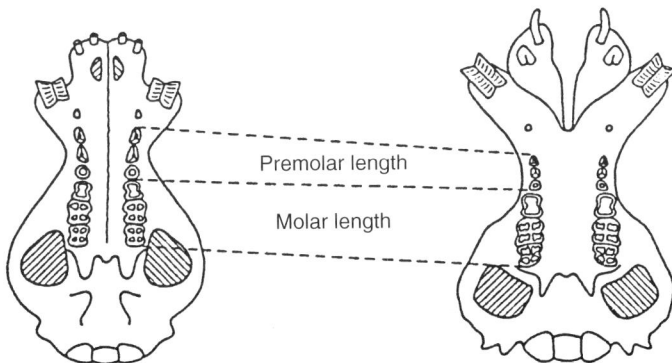

Figure 2.6 Diagram of the underside of the cranium of Pygmy hippopotamus (left) and Common hippopotamus (right) to show the different proportions of the premolar and molar lengths. From Coryndon (1978) in *Evolution of African Mammals* ed. by V.J. Maglio & H.B.S. Cooke. Copyright © 1978 by the President and Fellows of Harvard College. Reprinted by permission of Harvard University Press.

The full complement of teeth in the large hippo, therefore, is 32 in calves and 36 in adults although there may appear to be 40 with the retention of the fourth milk premolar. The pygmy hippo has only 34 permanent teeth but the calf is similar to its larger cousin in having 32 teeth.

Dental formula

The dentition of mammals is usually shown as a dental formula in which the numbers of teeth of each type are shown for half of each jaw. As the jaws are symmetrical, the total complement of teeth is twice the number shown in the formula.

The full dental formula for the family Hippopotamidae is:

$$\begin{array}{cccc} 3 & 1 & 4 & 3 \\ i & c\ pm & m \\ 3 & 1 & 4 & 3 \end{array}$$

but this is found only in ancestral fossil forms and in living hippos it becomes:

$$\begin{array}{cccc} 2 & 1 & 3\text{ or }4 & 3 \\ i & c & pm & m \\ 2 & 1 & 3\text{ or }4 & 3 \end{array}$$

for the large hippo and for *Hexaprotodon*:

$$\begin{array}{cccc} 2 & 1 & 3 & 3 \\ i & c & pm & m \\ 1 & 1 & 3 & 3 \end{array}$$

(Each line shows the number of teeth in one half of the upper or lower jaw. *i, c, pm* and *m* refer to incisors, canines, premolars and molars respectively.)

Tooth wear and age estimation

As in most mammals, the molariform teeth erode through use at a rate that can be used to estimate the relative age of the animal, the older the individual, the more worn down will be its teeth. The unworn tooth is covered with enamel but with use this is eroded to expose the underlying dentine. As the cusps of the teeth wear down, the dentine appears in a continuously changing pattern that reflects the age of the tooth. The order in which the teeth erupt from the jaw is also correlated with age.

Bill Longhurst, one of the Fulbright Scholars in Uganda (p. 89), devised a scale based on eruption and wear that allows a hippo to be assigned to a particular age group. The method can be used only with dead specimens whose lower jaws have been removed and cleaned of flesh. Longhurst's study was not published but Laws (1968a) refined the scale from an examination of 1244 lower jaws from the Queen Elizabeth National Park, Uganda. He converted the age classes to absolute ages in years on the assumption that the average life span of the Ugandan hippos is 41 ± 4 years. This estimate was based on the longevity of hippos in captivity but supporting evidence comes from a survey of hippos in Uganda; some animals were carrying anti-bodies to rinderpest, of which there had been an outbreak 42 years earlier. The oldest affected hippos – those with completely worn teeth – must have been infected some 42 years previously, suggesting that this period is approaching the maximum life span of the animals.

Laws' method of estimating the age of hippos from their teeth is widely accepted and is described below. The lower jaw is used as it is more convenient to handle than the skull. The length of the mandible obviously increases with age and is used as corroborating evidence although the ranges overlap in most cases as the growth rate slows down considerably in adults.

Laws uses 20 age classes, which are not of uniform length in terms of time. Some of the characteristics of each group are given in Table 2.1 and illustrated in Fig. 2.7. The eruption sequence of the milk teeth is used for the first few stages and their persistence can be used as markers up to Group VIII (11 years). Once the full set of permanent teeth has erupted, the age classification is based on the pattern of wear on the grinding teeth. The dentine is worn flat, or even concave, in old hippos and the patterns disappear. The canines and incisors are not much used in age estimation as sex differences in these teeth develop after Group VII (8 years).

Head

The head of *H. amphibius* is massive, with a jaw that is hinged very far back so that the gape is huge, approaching 180°. The skull is notable for the small brain case and the high supraorbital ridges, which are associated with the position of the eyes on top of the head. There is also a high sagittal crest for the attachment of the large temporalis

Table 2.1 Age classes determined from the molars (M) in the cleaned lower jaw of *Hippopotamus amphibius*. ("Exposed" means protruding above bone level, not erupted through the gum. Open means that the tooth is visible in its alveolus but not protruding above bone level. D = deciduous (milk) tooth, I = incisor, C = canine, P = premolar, M = molar.) (From Laws, 1968a)

Age Class	Age in Years		Mandible Length (mm)
I	0	D C,I1,I2.P1,P3 exposed. M1 open.	100–160
II	0.5	D C,I1,I2,P1–4 exposed. M1,M2 open.	182–260
III	1	D C,I present. C,I exposed. Wear on D P3,4. M2 open.	274–370
IV	3	D I1,2,3 lost. I1,2 exposed. C,M1 erupted.	318–398
V	4	Wear on C,I,M1. D M1 lost. P2,3,M2 open.	368–424
VI	7	D P1,2 lost. D P4 worn almost flat. M2 exposed.	410–466
VII	8	D P2,3 present. P2,3 erupted. M2 partly erupted. M3 open.	412–496
VIII	11	D M4 present. P2,3 slight wear, P4 exposed, M3 open.	454–516
IX	15	P2–4 erupted. M3 exposed. Wear on M2 on 2 cusps.	472–530
X	17	M3 erupted. Dentine on M1 usually continuous from wear.	462–536
XI	20	M1 dentine continuous. Slight wear on M3. Wear on P2,4.	496–544
XII	22	Increased wear P2–4. M1 worn almost flat.	496–588
XIII	24	M1 worn flat. M3 dentine exposed on 1st & 2nd cusps.	478–550
XIV	27	Further wear on all teeth. M2 cusps not yet joined.	482–570
XV	30	M1 flat/concave. 50% M2 cusps joined. 3rd cusp M3 worn.	506–566
XVI	33	M2 dentine continuous. Further wear on M3.	490–588
XVII	35	P2–4,M2 worn flat. M1 worn to gum. Continuous dentine M3.	510–602
XVIII	38	M1–3 worn flat.	510–574
XIX	40	M1 below gum, M2 concave, M3 flat. Resorption of bone.	490–556
XX	43	Usually only P3,4,M2,3 still present and very worn.	498–528

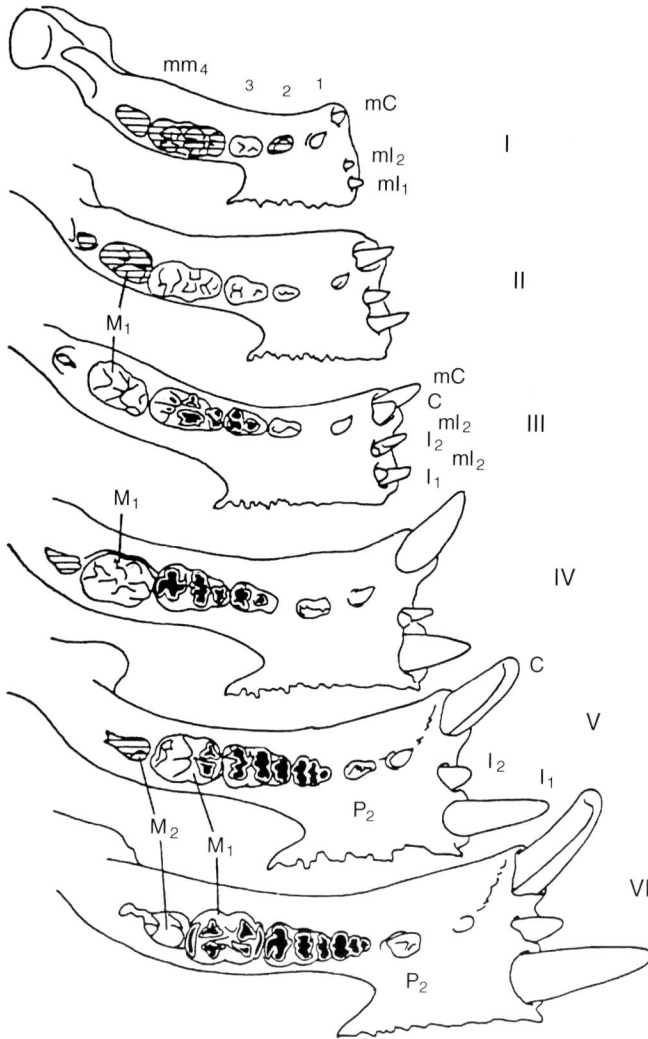

Figure 2.7 Age classes of the common hippopotamus based on the eruption and wear of the molars and premolars in the lower jaw. Black areas represent exposed dentine and a hatched area indicates an unerupted tooth within the alveolus below bone level. From Laws (1968a).

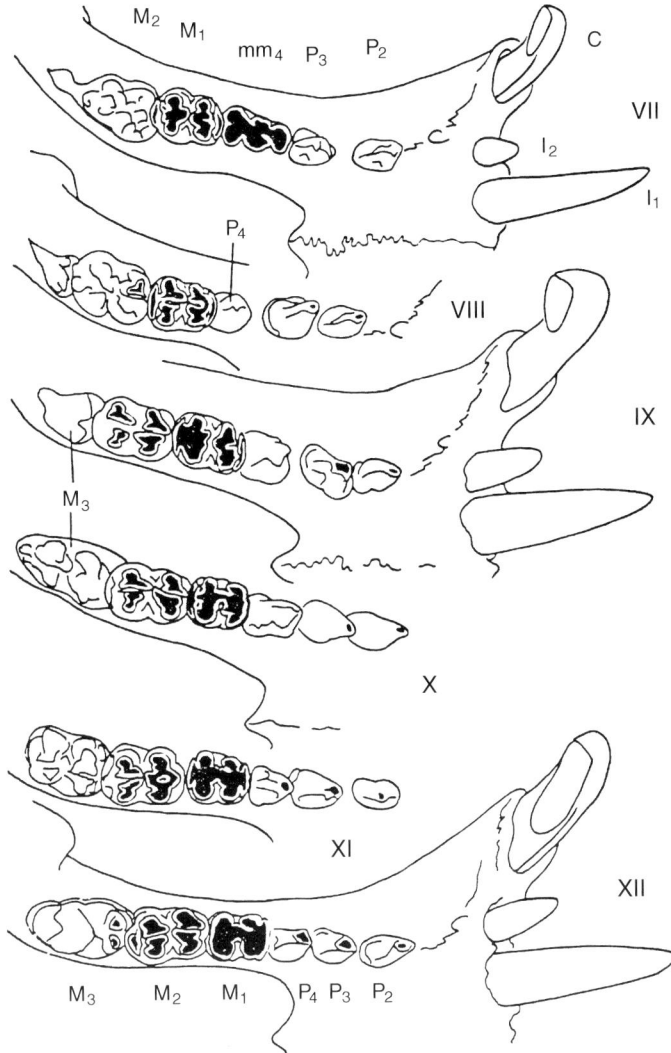

Figure 2.7 (cont.)

muscles, which pass through the massive zygomatic arch and, with the masseter muscles, control the movements of the lower jaw. These muscles are prominent in life on the back of the neck. The lower jaw is disproportionately large compared with most ungulates presumably because it has to act as a fighting weapon and carry the huge canines. The rear of the jaw is extended for the insertion of further powerful muscles and gives the hippo its marked jowls.

There are obvious differences in the shape of the skull in the two species. Viewed from below the skull of the large hippo is hour-glass-shaped while that of the pygmy hippo is more pear-shaped (Fig. 2.6). There are other technical differences which easily distinguish the two, e.g. in the size and position of the lachrymal and frontal bones. The most noticeable difference, however, is in the position of the orbits or eye sockets. These are high on the head in *Hippopotamus* and lie in the posterior half of the skull. Those of *Hexaprotodon* are found on the side of the head and are situated in a more midway position along the length of the skull.

Skin

The skin, which is very similar in both species, consists of a very thin epidermis overlying a much thicker dermis. Details of the fine structure of the skin of the common hippo are given by Luck & Wright (1964). The term pachyderm – thick skinned – is applied to the hippo with some justice for the skin is several centimetres thick over the back and hindquarters of the common hippo although it does thin to around a centimetre on the belly. It is an important organ, accounting for some 18% of the total body weight (Wright, 1964). Most of this weight is due to the dermis, which consists of a dense sheet of collagen with fibres arranged in a matted, but regular, pattern that gives great strength to the skin. Dispersed between the fibres is PAC-positive material, i.e. tissue that shows a positive reaction to periodic acid-Schiff reagents, a test that indicates the presence of specific carbohydrates. The other main component is a coarse network of blood vessels deep within the dermis. The thickness of the dermis varies with the position on the body. Table 2.2 shows some measurements made by Luck & Wright (1964) of the thickness at various distances from the middle of the back to the midline of the belly. It can be seen that the skin thins abruptly once the maximum width of the body is passed. The skin is also thinner on the inner surfaces of the limbs and is more pliable there as well as over the belly.

Collagen is a gelatinous protein that occurs as connective tissue in many animals. In the hippo, it consists of 64% water and 1% lipid material (fat). When removed from a dead hippo and oiled, the collagen becomes translucent and can be used as window glass. Dried skins also make useful battle shields.

The upper, cornified layer of the skin, the epidermis, is very thin and, being well supplied with nerve endings, is said to be very sensitive. It needs to be kept moist for it cracks easily when dry. The epidermis of the common hippo is from 0.5 to 0.9 mm in thickness over the abdomen and from 0.7 to 1.0 mm on the back. The boundary between the epidermis and dermis is characterised by deep epidermal ridges with

Table 2.2 The thickness of the skin of the common hippopotamus at various distances from the midline of the back. The nominal girth is 300 cm so that the distance extends as far as the midpoint of the belly. (After Luck & Wright, 1964).

Distance from midline of back (cm)	Thickness of Skin (mm)
0	22
20	35
80	35
90	12
150	12

corresponding dermal papillae protruding from the dermis at a linear density of seven to nine papillae per millimetre.

The most obvious feature of the skin is that it appears hairless apart from a few bristles around the mouth and on the tail. There is, however, a covering of fine hairs at a density of 20–30 per 100 cm^2 over the back, with about half that number on the flanks and even fewer over the belly. There are no true sweat (sebaceous) glands either within the dermis or at the base of the few hairs, where such glands are almost always found in most other mammals. There are, however, large subdermal glands, made up of more than one type of cell, scattered regularly over the body (Wright, 1987). They are sparsely distributed at a density, in the common hippo, of only one per cm^2 but they make up for that by being unusually large for a skin gland, with a weight of about 1 g and a duct clearly visible to the naked eye. Hence the volume of secretion is quite high.

These glands produce a viscous fluid varying from colourless to pink and dark brown. It seems likely that the secretion turns brown on exposure to air and the brown patches around the ducts are most probably dried secretion. The coloured secretion gives a pinkish tinge to the whole body. The fluid is highly alkaline with a pH range of 8.5 to 10.5 in *H. amphibius* and 9.5 in the pygmy hippo (Olivier, 1975). Wright (1987) analysed the ionic composition of the secretion collected from a tame hippo at 2 o'clock in the afternoon under a complete cloud cover and obtained the following results:

$$Na^+ \ 1.2 \ mmol \ dm^{-3}$$
$$K^+ \ 140.0 \ mmol \ dm^{-3}$$
$$Cl^- \ 18.4 \ mmol \ dm^{-3}$$
$$HCO_3^- \ 99.5 \ mmol \ dm^{-3}$$

The composition of the fluid varies over time but in a way which is not clear (Wright, 1964). It is thought to have antiseptic properties and to play a part in preventing the infection of wounds. Certainly, serious wounds sustained in fights seem to clear up remarkably quickly despite the foul surroundings of a hippo wallow.

Olivier (1975) analysed the secretion from a pair of pygmy hippos held at Bristol Zoo and found that the concentrations of potassium (K$^+$) and sodium (Na$^+$) ions were much lower than those reported for the common hippo by Luck & Wright (1964). The relevant measurements (in mEq/l) for each species (*Hexaprotodon* first) were 166.0 and 283 for K$^+$ and 3.6 and 18 for Na$^+$.

The skin colour of the large hippo is greyish black, running to pink over the abdomen and around the eyes and ears. Sometimes the secretion from the dermal glands gives a pinkish tinge to the whole body. The pygmy hippo is more uniformly black with no pink coloration. There have been some reports of albino hippos by Captain Pitman (1962), mainly from Uganda. One, sighted in March on the Nile downstream from the Murchison Falls, was a "hideous" pink colour with liver-brown spots sprinkled over its head and flanks. Several other sightings of pink hippos were made in the Queen Elizabeth and Murchison Falls National Parks in the 1950s and another was seen in the Kagera River, which forms the international boundary between Tanzania and Uganda. Those in the Murchison Falls National Park were said to be unusually timid. The only reported instance of pink hippos away from Uganda is of two seen by Mr W.F. Stubbs in Gorongosa Game Reserve in Mozambique. One was standing up to its belly in turbid water but the exposed skin was all pink. Another, half a mile away, lacked pigmentation on one side from its face down to its forequarters, giving it the appearance of a Hereford steer. None of these animals was a true albino as apart from possessing some brown coloration, their eyes were not pink. The pink colour of the body was probably due to the secretions of the subdermal glands, possibly augmented by the network of blood vessels under the skin. The main problem for an albino hippo is presumably sunburn but all the pink hippos seen were adult and apparently healthy.

Digestive System

Large mammalian herbivores tend to fall into two categories, the fore-gut digesters, in which the food is digested in the stomach, and the hind-gut digesters, in which digestion takes place in the caecum and/or the large intestine (as these organs are embryologically derived from the mid-gut, such herbivores should really be called mid-gut digesters). Alternative names are gastric and post-gastric digesters, depending on whether or not digestion takes place in the stomach or posterior to it.

The digestive system of hippos is of great interest in that they have a stomach very similar to that of a ruminant. Ruminants differ from other herbivores in that they have a chambered stomach, unlike the simple single stomach of, say, a horse. They also differ in being fore-gut digesters. There are four compartments in the bovid stomach – the rumen, reticulum, omasum and abomasum, listed in the sequence in which food passes from one compartment to the other. The abomasum is the only portion of the stomach to secrete gastric juices and hence is sometimes called the "true" or gastric stomach. Cervids (deer) and giraffes also have four-chambered stomachs but camels have only three chambers as do water chevrotains, a primitive group of artiodactyls with similarities to deer, antelopes and pigs.

All of these herbivores ruminate, i.e. chew the cud, but there are other groups which do not ruminate yet have chambered stomachs. These are sometimes called pseudo-ruminants. They include the hippo, which has perhaps the most complex stomach of them all. One can trace a sequence in the pseudo-ruminant condition starting with the pig, whose stomach, although simple, is divided into glandular and non-glandular portions. Next along the line are the peccaries (Tayassuidae), which have a two- to three-chambered stomach. The hippo is said variously to have a three- or four-chambered stomach according to which text-book one reads. Champions of three chambers include Nowak & Paradiso (1983), Anderson & Jones (1984), Grzimek (1988) and Skinner & Smithers (1990) but authors who have actually dissected a carcase describe four compartments. I suppose it depends on what constitutes a chamber for it is possible that the first two stomachs could be considered as one that has split into two and that a true rumen is absent.

The gut anatomy of the two hippo species are very similar although the dimensions are, of course, different. They will be considered separately here.

The stomach and intestines of *Hippopotamus amphibius*

One of the first descriptions of the gut anatomy of a hippo is that of Crisp (1867). The hippo in question was not full grown, being only 14 months old. It had been burned to death in a fire at London's Crystal Palace, where it had presumably been exhibited (this was not the big fire which destroyed the Palace in 1936). Crisp mentions that one side of the animal was "well roasted" and together with some friends, he sampled the flesh and pronounced the flavour to be "excellent" and the meat to be "whiter than any veal I have ever seen". This must have been one of the earliest hippo barbeques and probably the first in England.

Apart from the culinary tips, Crisp gives a detailed drawing of the stomach (Fig. 2.8) which clearly shows four compartments. The first two form equal-sized diverticula at the point where the oesophagus ends. The right chamber has a muscular, crescent-shaped structure which Crisp suspects serves to direct the food into the left

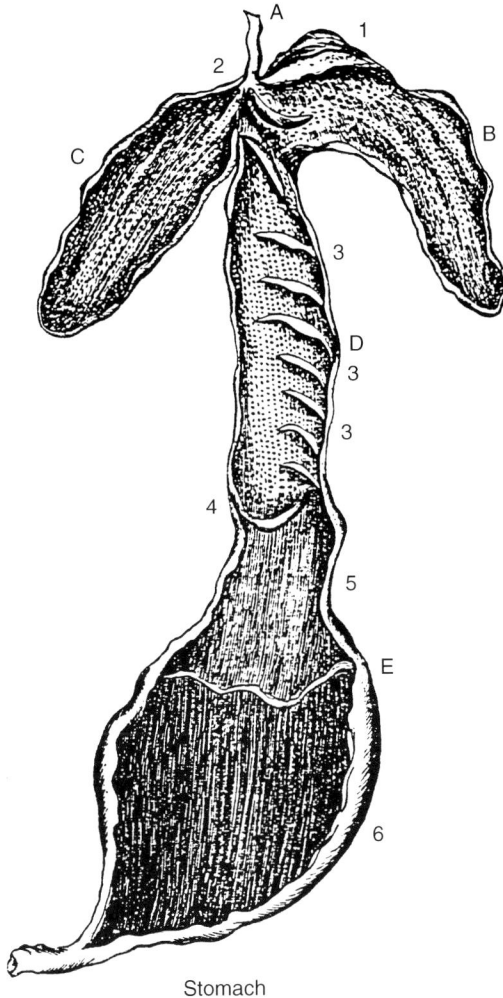

Stomach

Figure 2.8 Stomach of the common hippo according to Crisp (1867). Key: A – oesophagus; B, C – "first and second stomachs"; D – third stomach; E – glandular stomach; 1 – muscular lip directing food into the left stomach; 2 – valvular projection directing food into third stomach; 3 – valves; 4 – crescent-shaped valve; numbers 5 and 6 are not explained.

stomach, which would, therefore, be analogous to the rumen of a ruminant. The right diverticulum, however, is larger in the adult according to a footnote in Crisp which describes a dissection by a Dr Peters, who found it to be twice the length of the left. This suggests that it is the right diverticulum that is equivalent to the rumen and which first receives the food. The lining is similar to that of a rumen with about 65 longitudinal rows of rounded papillae. The third chamber is cylindrical in shape and also lined with papillae similar to those found in ruminants. Seven valve-like flaps project into its cavity and possibly serve to slow down the rate at which the food passes through the stomach. The lining of the fourth compartment is quite smooth and lacks papillae. It appears to be glandular, like the fourth stomach of a ruminant. The anterior end is white and differs in appearance from the posterior region, which is red and elevated, suggesting differing functions. Crisp (1867) gives some figures for the volume of the stomach's compartments and although the actual values are of limited significance due to the hippo's immaturity, the comparative figures may be of interest (Table 2.3).

Over a century later, Arman & Field (1973) dissected the stomach of a hippo foetus and came to much the same conclusions as Crisp. They described two anterior diverticula followed by a large median chamber leading to a posterior compartment. The left diverticulum is divided by a septum which is probably the crescent-shaped structure described by Crisp. The first three chambers are lined with papillae, which are larger in the two diverticula, but the lining of the fourth compartment is glandular. A muscular tract runs from the oesophagus to the posterior chamber and may be equivalent to the *sulcus reticuli*, the reticular groove found in ruminants and considered by Black & Sharkey (1970) to be essential for ruminant digestion. The groove, however, is very much bigger in the hippo than in ruminants. Arman & Field (1973) assume that fermentative digestion occurs in the first three compartments with gastric digestion in the fourth. They suggest that the probable route of the food is from the oesophagus to the diverticula and then on to the median chamber for sorting, with some finer particles passing back into the diverticula for further fermentation. They estimated relative quantities of the food in the various chambers to be 35% in the diverticula, 60% in the median chamber and 5% in the posterior chamber. They remark on the dryness of the stomach contents compared with those of a typical ruminant and suggest that conditions are less favourable for fermentation and absorption.

Van Hoven (1978) provides a schematic diagram of the hippo stomach following the terminology used by Langer (1976). This structure is based on 69 animals culled in the Kruger National Park in 1976 (Fig. 2.9). He recognises four, if not five, chambers with two blind sacs, a visceral and a parietal, each divided into a dorsal and ven-

Table 2.3 Volumes and dimensions of the stomach chambers of a 14-month-old hippo. (After Crisp, 1867.)

Chamber	Capacity (litres)	Dimensions (cm)
Right diverticulum	3.41 (20%)	48 × 15
Left Diverticulum	3.41 (20%)	53 × 11
Third Chamber	3.41 (20%)	51 × 13
Fourth Chamber	6.82 (40%)	36 × 25

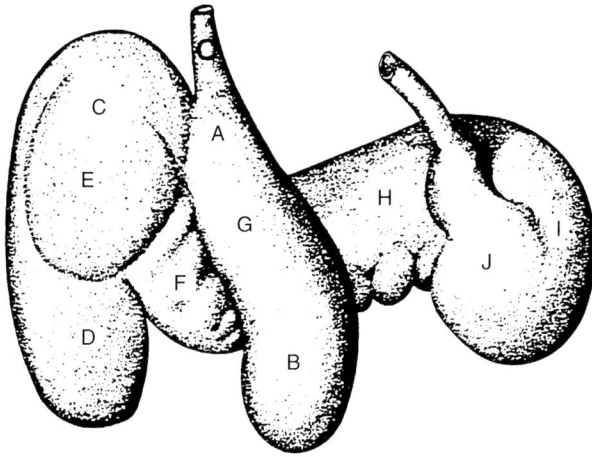

Figure 2.9 Schematic diagram of the stomach of a common hippopotamus. Key: O – oesophagus; A,B – visceral blind sac; C,D – parietal blind sac; E – vestibulum; F–I – connecting chamber; J – glandular stomach. The blind sacs are equivalent to Crisp's first and second stomachs and the connecting chamber is Crisp's third stomach. After Van Hoven (1978).

tral portion, a vestibule and a connecting chamber leading into the gastric stomach. He states that the food passes from the oesophagus to the visceral blind sac, then via the parietal blind sac to the vestibule. From there it passes along the connecting chamber to the gastric stomach.

Clemens & Maloiy (1982), who studied the digestive physiology of three species of pachyderms, give some information on the relative capacities of the stomach compartments and the rest of the gut from a *post mortem* examination of an adult hippo from Kenya. Their terminology follows that of Van Hoven (1978) but although the visceral blind sac could be clearly defined, they found that accurate demarcation between the parietal blind sac, vestibule and connecting chamber was not possible. Consequently they numbered some compartments as Stomachs 2 to 3. Stomach 2 was predominantly the parietal blind sac and Stomach 3 the proximal part of the connecting chamber. The vestibule was divided between Stomachs 2 and 3. Stomach 4 was the distal part of the connecting chamber. As they were interested in the progressive digestion of food they were not too concerned over the precise terminology of the stomach's anatomy. They did, however, provide some information on the contents of the stomach, and their results, given in Table 2.4, should be compared with those in Table 2.3.

The alimentary canal posterior to the stomach is unexceptional. There is no caecum and the large and small intestines are similar in external appearance. It is generally assumed in the literature that there is no gall bladder, despite the mention of one in the postscript to Crisp's 1867 paper, but a recent study has confirmed its presence (Gansberger & Forstenpointner, 1995). A large gland, some 15 cm long and covering an area of about 40 cm^2, lines the inner surface of the intestine where it widens into the colon. Similar, but smaller, glands are found in the jejunum and ileum.

Table 2.4 Weight of contents of the stomach chambers and intestines of an adult hippopotamus shot in Kenya. (From Clemens & Maloiy, 1982.)

Chamber	Wet Weight of Contents (kg)		Dry Weight of Contents (kg)	
Stomach blind sac	29	(15.5%)	15.0	(13.2%)
Stomach 2	38	(20.3%)	15.8	(13.0%)
Stomach 3	50	(26.7%)	16.4	(14.4%)
Stomach 4	21	(11.2%)	13.0	(11.4%)
Glandular stomach	3	(1.6%)	8.6	(7.6%)
Small intestine	27	(14.4%)	24.6	(21.6%)
Colon	19	(10.2%)	20.5	(18.0%)

Accessory organs include the pancreas, which is three-lobed, a long, narrow spleen and a uni-lobular liver, although the latter could be said to be lobular as there are several clefts in its surface.

Crisp (1867) gives some dimensions for the lengths of different parts of the gut and although these refer to only one immature male, their relative lengths may be representative of the species (Table 2.5). The length of the gut of a full-grown hippo was estimated to be 42 m and extrapolations to this length are included in the table.

Table 2.5 Length (metres) of each section of the gut in a young male hippo. Equivalent values for a full-grown hippo are included assuming that the total length in the latter is 42 m. (After Crisp, 1867.)

Organ	Actual Length	Extrapolated Length
Oesophagus	0.76	0.89
Stomach	1.78	2.08
Small Intestines	30.38	35.43
Large Intestines	3.09	3.60
Total Length	36.01	42.00

The stomach and intestines of *Hexaprotodon liberiensis*

There have been only a few studies of the gut anatomy of the pygmy hippo, the latest of which is that by Macdonald & Hartman (1983), who dissected a 12-year-old male and a stillborn female. There are four chambers comprising two blind-ended sacs, a connecting chamber and a glandular stomach (Fig. 2.10). This is essentially the same condition as that found in the large hippo. The oesophagus leads into the visceral blind sac, which, in turn, leads into the parietal blind sac. This is linked to the glandular stomach by the connecting chamber, which has nine semi-lunar flaps or folds rather than the seven reported for the large hippo. This is probably the route, shown in the figure, whereby the food passes through the stomach chambers. The first three chambers are non-glandular and are lined

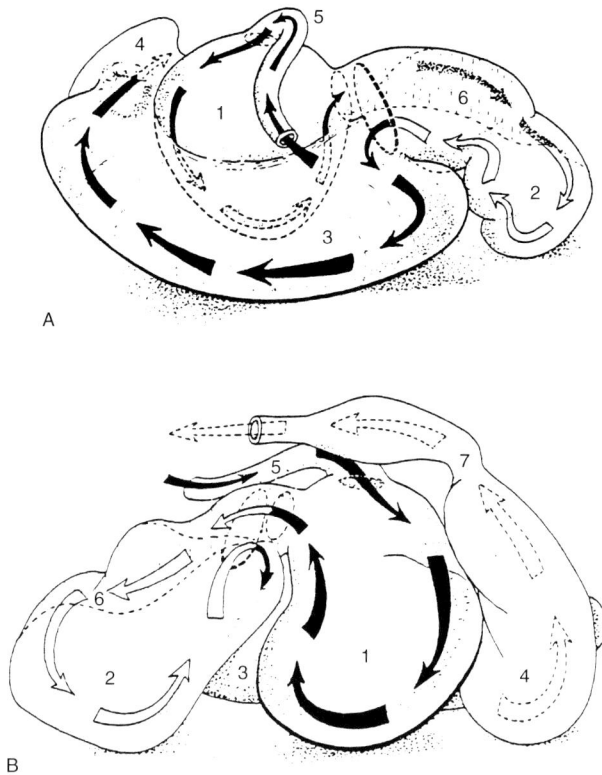

Figure 2.10 Schematic representation of the stomach of a pygmy hippopotamus seen from the front (A) and rear (B). Key: 1 – visceral blind sac; 2 – parietal blind sac; 3 – connecting chamber; 4 – glandular stomach; 5 – oesophagus; 6 – longitudinal fold in parietal blind sac; 7 – pyloris. Arrows show the route of food through the chambers. A comparison with Fig. 2.9 shows that the structure of the stomach in the two species is fundamentally the same. From Macdonald, A.A. & Hartman, W. (1983) Comparative and functional morphology of the stomach in the adult and newborn pygmy hippopotamus (*Choeropsis liberiensis*) *Journal of Morphology*, **177**: 269–276. Reprinted by permission of Wiley-Liss, Inc., a subsidiary of John Wiley & Sons Inc.

with mucous membrane. Fermentative digestion presumably takes place in the blind sacs.

The orientation of the connecting chamber was found to be radically different in the new-born hippo compared with that in the adult. In the latter, the chamber lies transversely and connects with the glandular stomach on the right side. In the new-born, however, it lies vertically and connects ventrally with the glandular stomach. Macdonald & Hartman (1983) point out that the illustration given by Arman & Field (1973) of a foetal *H. amphibius* also seems to show the connecting chamber in a vertical orientation and that Langer's descriptions point to a horizontal arrangement in the adult. It seems likely therefore, that in this aspect, as in others, the gut morphology of the two species is remarkably similar. Macdonald & Hartman (1983) suggest that the difference could be associated with lactation. There is a groove in the visceral sac and the connecting chamber that perhaps is closed off to form a tube through which milk can pass straight into the glandular stomach. Such an arrangement is found in ruminants in which the closure of the tube is a reflex action caused by sucking.

The intestines of the pygmy hippo form a simple tube without a caecum, as in the large species. The junction between the small and large intestines is marked by an increase in diameter from 2.5 to 5.0 cm and a change in the gut lining from a smooth to a wrinkled appearance. The length of the small intestine (duodenum, jejunum and ileum) is about 16 m and that of the large intestine (colon and rectum) is about 2.5 m. The relative lengths of the two intestines is similar in the two hippo species.

Large amounts of fat were found in the mesenteries of the small and large intestines in the newborn animal. Macdonald & Hartman deduce from this that the energy metabolism of the hippo is radically different from that of their supposed relatives, the pigs, whose new-borns store energy as glycogen, but is similar to that of ruminants, in which fat is also the main energy store for neonates.

Reproductive Organs of Hippopotamus amphibius

The anatomy of the reproductive organs – male

The testes of mammals develop within the abdomen but they usually descend into a scrotum not long before birth. The traditional explanation for this is that the body temperature is too high for spermatogenesis to occur but this does not appear to be a problem with some species, such as elephants, in which the testes remain abdominal throughout life. The hippo is intermediate for although the testes descend, they do so only partially and are not contained within a scrotum (Laws & Clough, 1966). The epididymis, which is the organ within which the spermatozoa mature and are stored before ejaculation, lies along the posterior surface of the testis and is divided into the head, body and tail. In the hippo, the head appears to have an absorptive function whereas the tail, with its large capacity, is a storage organ for semen (Kayanja, 1989; Muwazi & Kayanja, 1991). The body appears to have an intermediate function.

Fine detail of the histology of the testis and its associated ducts is given by Kayanja (1989). The interstitial tissue of the testes differs from that in other Suiformes, i.e. the pigs and peccaries, in being unpigmented. The reproductive tract conforms to the basic mammalian plan, with paired seminal vesicles, which contribute to the seminal fluid, and paired Cowper's glands. These also provide secretions for the semen as does the prostate gland, which is of the disseminated type. The bulbo-urethral gland, which provides fluid for the semen, is remarkable for its large size. The fibro-elastic penis is a large S-shaped organ with paired retractor ligaments attached to the point of flexure. As the penis is normally retracted and there is no scrotum, it is not easy to determine the sex of a hippo at first glance. Some measurements of the male reproductive organs are given in Table 2.6.

At least two of the bulls examined by Kayanja were territorial males and the others were shot either in wallows or on land at night but all were fully fertile.

The anatomy of the reproductive organs – female

The female reproductive tract has been described by Laws & Clough (1966). It does not differ markedly from that of other eutherian mammals except in two particulars,

Table 2.6 Average weights and measures of the reproductive organs of ten adult male hippos culled in Queen Elizabeth National Park, Uganda. (From Kayanja, 1989.)

Testis weight	275 g
Epididymis weight	135 g
Bulbo-urethral gland weight	110 g
Bulbo-urethral gland diameter	8 cm
Penis weight	1616 g
Penis length	111 cm

mentioned below. The paired ovaries are partially enclosed in membranes and both may be active in producing eggs. Of the 97 hippos examined by Laws & Clough, the corpus luteum of pregnancy was in the left ovary on 50 occasions and in the right on 47, a difference that is not statistically significant. It is likely that ovulation occurs alternately between the ovaries of an individual. The fallopian tubes, which lead from the ovaries to the uterus, are short, with an extended length averaging only 42.5 cm in five dissected animals. The uterus is Y-shaped (bicornuate) with a very short base so that the fertilised egg embeds in one or other of the two uterine horns. Presumably the ovum embeds in the horn leading from the ovulating ovary. The hymen is imperforate in young animals and persists as a small hymenal fold at the junction between the vagina and vestibule. The clitoris is large at 6.5 cm in length.

The unusual features of the female reproductive tract concern the vagina and its entrance, the vestibule. The upper part of the vagina is marked by a series of transverse, interlocking fibrous ridges numbering from ten to 19, large in front but decreasing in size posteriorly. Their function is unclear. Somewhat similar ridges have been noted in pigs, particularly in the warthog, although their ridges are less pronounced. Smaller, spiral ridges in the vagina of the domestic pig match the corkscrew shape of the boar's penis but there is no evidence, as far as I know, that the penis of the hippo locks into the female ridges. Interestingly enough in view of the DNA similarities mentioned on page 45, transverse ridges have also been reported in cetaceans, although they are less prominent than in the hippo. An aquatic adaptation has been postulated for these ridges in the cetaceans although it is difficult to see what that could be.

The second peculiarity of the female hippo's reproductive tract is the presence of two large diverticula projecting from the vestibule. The function of these structures is unknown.

A curious anomaly was noted by Clough (1970) in one of the female hippos he examined. The ovary of this individual, which was about 31 years old and lactating, contained tubules similar to the seminiferous tubules produced in the embryonic testis. Clough was unable to explain the origin of these testis cords in an ovary but the animal was fertile and not a true hermaphrodite. Similar structures have been found in the ovaries of other species but this is the only case reported for a hippo.

Smuts & Whyte (1981) found a probable case of a teratoma in an ovary from a hippo shot in the Kruger National Park, South Africa. A teratoma is a neoplasm, usually in the testis or ovary, composed of different types of tissue from that of the organ concerned. In the present case, the atypical tissues were columnar with underlying mucus cells usually associated with respiratory mucosa. There were also spicules of cartilage present. The abnormal ovary was, at a weight of 687 g, much larger than the other one, which weighed a normal 54.5 g.

Another of the hippos examined by Smuts & Whyte had undeveloped ovaries, with a combined weight of only 2.5 g. She was at least 11 years old so was fully mature physically. The ovaries consisted mostly of muscle fibres and there were no follicles. It is possible that the blood supply to the ovaries had been impaired for there were lesions in the blood vessels caused by parasitic blood flukes.

The placenta of the hippo is similar to that of the pig and is technically epitheliochorial and of the diffuse type.

Physiology

The physiology of the hippo is distinctive mainly in the skin. The physiologies of digestion and reproduction are touched on in the relevant chapters but as they do not differ markedly from those in other large mammals, they will not be considered further here except to emphasise that the salient feature of the digestive physiology of the hippo is the near ruminant function. Although the condition is not unique, the hippo probably has the most complicated chambered stomach to be found in a non-ruminant.

Skin physiology

The principal physiological function of the skin is to control body temperature through the regulation of water loss. Wright (1987) calculated the water loss from the large hippo by measuring the rate of change in humidity within a capsule applied to the skin of a tame animal relative to the humidity in the surrounding air. Evaporation from the hippo's skin was found to be very high compared with that from other mammals and varied from 68 g $m^{-2}h^{-1}$ when the skin was dry, to 2280 g $m^{-2}h^{-1}$ when the skin was wet with secretion (Wright, 1987). The latter value approaches the figures obtained by placing the hygrometric capsule over filter paper soaked in a KCl solution, suggesting that there is no control over water loss from the skin. One inference that could be drawn from this is that temperature control is not achieved through any mechanism similar to sweating unless the red secretion acts as sweat.

Wright and Luck measured the evaporation from the skin when the subdermal glands were not active (Wright 1964) and found that the rate of loss was proportionately several times that from a man under similar circumstances. They also found that evaporation was increased when there was a layer of dried secretion present. They attribute the high rate of water loss to intercellular material in the skin, which acts almost like a wick in transferring water from the body to the air. If, as this might suggest, the hippo is unable to control the rate of water loss from its body, its ability to deal with heat stress will be severely limited. It may be for this reason that the hippo has evolved a semi-aquatic habit, in which temperature control is achieved by spending the day in the water, where temperature is relatively constant. The same is true of

Table 2.7 Insensible water loss from the skin of the two hippo species. (After Olivier, 1975.)

Specimen	Evaporation in mg/5 cm²/10 min	
	Mean	*Range*
Dead *H. amphibius*	15.9	9.1–24.3
Living *H. amphibius*	12.7	7.2–22.4
Living *H.liberiensis*	5.7	3.0–11.2
H. liberiensis corrected for temperature	11.4	6.0–22.4

the air temperature at night and again, this may be a contributing cause of the hippo's nocturnal behaviour.

Olivier (1975) measured water loss from the skin of tame pygmy hippos and compared them with data available at the time from the large hippo (it is important that measurements should not be taken when the animal is "sweating"). The results, which were expressed in different units from those mentioned above, are given in Table 2.7. They reveal a much lower water loss in the smaller species; about half that in the larger. This is surprising given the close similarity in the skin anatomy of the two species, but the temperature was some 10°C higher during the measurements of the large hippo and as the water vapour pressure doubles with each ten degree rise, the water loss would also double. Thus, if the values for the pygmy hippo are doubled, they come within the range of the values for the large species. This seems more reasonable and suggests that the physiology of the skin as well as its anatomy is almost identical in the two species.

The circumstances under which the red fluid is secreted are not clear. Secretion has been observed on the skin under water, where presumably it can have no cooling function (Wright, 1964). It is produced on land during the day but not usually at night, which is consistent with the hypothesis that the fluid functions as sweat. Verheyen (1954) maintains that the secretion is particularly copious in fighting males and in females about to give birth.

Body temperature

The hippo may or may not use its skin secretion as sweat but whatever the mechanism may be, the deep body temperature is maintained at a constant level. This was demonstrated by Wright (1987), who placed a thermistor some 300 mm into the rectum of his tame *H. amphibius* and kept it there for the nine hours of observation. The results showed that the core temperature was maintained throughout the day to within one degree of 36°C under a variety of environmental conditions that were probably harsher than those obtaining in the wild.

This is close to the rectal temperatures taken by Luck & Wright (1959) soon after death from wild hippos culled in the Queen Elizabeth National Park, Uganda. Their measurements were spread over the 24 hours and were taken six months apart from two groups of hippo; some had been wallowing and some had been shot on land. Temperatures were recorded in Fahrenheit with a mean of 95.7°F (±0.7°), which is equivalent to 35.4°C. Hippos that had been shot in wallows had rectal temperatures that were usually below this figure whereas those on land showed a value slightly above average but there was no obvious variation with time of day. The skin temperatures on the back and belly were also measured and it was found that those of wallowing hippos were considerably lower than the core temperature. The temperature of unsubmerged back skin rose during the day until it was equal to or greater than the rectal temperature. There was little difference between the temperatures of back and belly skin in animals that had been shot at night or in hippos that had been completely submerged before death. Whatever the skin temperature may be, the core temperature is reached within 70 to 150 mm from the surface, depending on the position on the hippo's body.

Luck & Wright made some further measurements of the core temperature of hippos from the same region extending over four years and they found that the mean still clustered around the 35.4°C mark with a standard deviation of only 0.4°C (Wright, 1964). Some results are given in Fig. 2.11, which shows that the core temperature does not vary much despite wide changes in environmental conditions and confirms the tendency noted above for animals shot on land to have higher body temperatures than those shot in the water.

The constancy of the deep core temperature in the hippo contrasts with that of some other herbivores of similar size such as the white rhinoceros (Allbrook *et al.*, 1958). The rectal temperature of a captive rhino rose during the day from 34.5°C at sunrise to 37.5°C at sunset. When kept in the shade all day the rectal temperature remained constant at 35.2°C. The air temperature was 25°C with a relative humidity

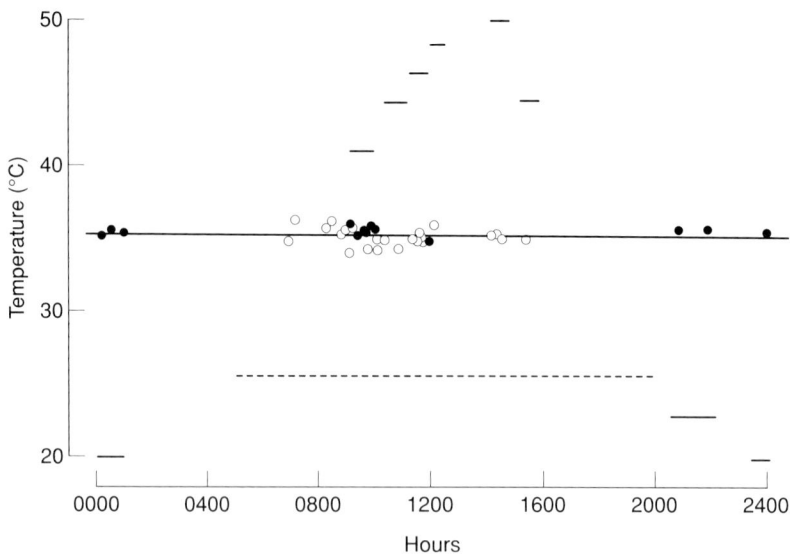

Figure 2.11 Deep body temperature of common hippos shot or immobilised at various times of day. Black circle – hippo taken on land, open circle – taken in the water. The horizontal line shows the mean temperature of all the hippos. Short horizontal lines represent the mean air temperature and the dashed line, the mean water temperature at various times when measured. From Wright (1964).

of 75%. At night when the air temperature fell to 5°C, the rhino's core temperature was only 33.6°C.

The difference between the two species reflects their different life styles. The rhino is similar to many other grassland mammals in allowing the deep body temperature to change throughout the day. For a land mammal, the maintenance of a constant body temperature requires the expenditure of energy and water, which the animal can scarce afford to waste. Heat is bound to be absorbed from the sun but instead of trying to off-load it through sweating or panting, the animal can absorb it by allowing the body temperature to rise. Nights are usually much cooler and the absorbed heat can then be dissipated into the atmosphere. This ploy is not necessary in the hippo as it can control its temperature to a large extent by moving onto land if it is getting too cold or back into the water if it is over-heating. The mean water temperature in the wallows during the observations on the Ugandan hippos ranged from 23.8–26.1°C, well below body temperature and therefore suitable for cooling purposes.

Digestive physiology

Plant leaves, particularly those of grass, the principal food of hippos, contain a high proportion of cellulose, which large mammals cannot digest by themselves but have to rely on micro-organisms in the gut to ferment it for them. The ruminant's strategy is to retain the food in the stomach for a long period and subject it to thorough digestion. This takes time, which is worth spending only if the food is of high quality. An alternative strategy is to pass the food through quickly, as the horse does. This is the more efficient system with poor quality food but it does mean that large quantities have to be eaten, much of which is not properly digested. The ruminant has the advantage with food of good quality for apart from extracting more nutritive value, it has a longer stretch of intestine, as a fore-gut digester, for the assimilation of digestive products than is the case with a postgastric digester.

The function of chewing the cud is to break down the material into small particles which can then be processed more efficiently by the micro-organisms in the gut. The hippo does not chew the cud but its chambered stomach suggests that it has a ruminant-like digestion, which is why it has been termed a "pseudo-ruminant". It is likely that the arrangement improves the digestive efficiency of the animal although this has not been established experimentally.

The chambered nature of the hippo's stomach led to the assumption (Moir, 1965) that it also experienced fermentative digestion but there was little direct evidence. Circumstantial evidence came from the discovery by Dinnik *et al.* (1963) of ciliates in the stomach. Fermentation was confirmed by Thurston *et al.* (1968), who demonstrated high concentrations of volatile fatty acids in the stomach together with a rich and diverse ciliate fauna. Further evidence was provided by Duncan & Garton (1968), who found that the fatty acid composition of fat deposits was similar to that of ruminants. There are concentrations of volatile fatty acids in the intestines suggesting that some fermentative digestion occurs there.

The study of hippo digestive physiology by Clemens & Maloiy (1982) showed conclusively that the hippo ferments its food, primarily in the chambers of the fore-stomach. Thus the concentrations of volatile fatty acids, the breakdown product of digestion, were much higher in the stomach chambers of the hippo than in the

Table 2.8   Concentrations (mmoles per litre) of volatile fatty acids in the gut contents of hippos, elephants and black rhinos. High values indicate regions where most digestion occurs. The small intestine and colon were divided into sections for analysis. (After Clemens & Maloiy, 1982.)

Region of gut (Elephant & Rhino)	Elephant	Rhino	Hippo	Region of Gut (Hippo)
			150.4	Blind sac
			153.3	Stomach 2
Cranial Stomach	10.3	34.9	140.0	Stomach 3
			60.0	Stomach 4
Caudal Stomach	9.7	42.6	30.1	Gastric Stomach
Small Intestine 1	10.1	37.3	37.7	Small Intestine 1
Small Intestine 2	12.8	39.6	28.3	Small Intestine 2
Small Intestine 3	17.1	51.7	48.5	Small Intestine 3
Caecum	137.6	144.4		
Colon 1	121.9	79.5	28.0	Colon 1
Colon 2	128.7	80.7	34.9	Colon 2
Colon 3	148.3	78.6		
Colon 4	114.2	72.0		
Colon 5	64.6	53.3		

stomachs of the elephant and rhino. The reverse was true farther down the gut showing that most digestion takes place in the colon and caecum of the non–ruminant (Table 2.8).

Digestibility of food

The digestibility of food can be roughly calculated by comparing food intake with faecal output, e.g. a measure of the amount of protein that has been digested is the difference between the protein content of the food and the amount of protein in the droppings. It is not easy to make such measurements with hippos because of their habit of defaecating in the water, which makes collection difficult if a day's production is to be analysed. This can be circumvented by keeping the animal out of water but then the skins dries out and the experimental subject loses condition, which, apart from animal welfare considerations, invalidates the observations. Abaturov *et al.* (1995) studied digestion in hippos but are rather vague about techniques. They needed to know the daily faecal production for their formulae, and presumably calculated this by extrapolation from the weight of a single defaecation and an estimate of the number of defaecations in a day. From these they estimated the proportion of each component of the food that was digested and calculated the digestibility coefficient (D), which is given by the equation

$$D = 100(1-v/f)$$

where v and f are the lignin concentrations in the food and faeces respectively. Lignin is taken as the control as no mammal is able to digest it.

The results of their analyses of three hippos are given in Table 2.9. The percentage of crude protein in the diet, a popular if imperfect measure of food quality, is nearly 50% higher in hippos than in the two other grazers, zebra and hartebeest, included in the study. This probably reflects the hippos feeding style, which ensures a continuous supply of young, high quality grass. The digestibility of dry matter by hippos (45%) and zebra (42%) was much lower than for hartebeest (51%) or Grant's gazelle (61%). The two last species are ruminants, which, as we have seen, are more efficient digesters of high quality food than monogastric ungulates. The gazelle is a browser as well as a grazer and the quality of its food is therefore higher, e.g. the crude protein level was 18.09% in this study. The digestibility coefficients of other components in the food was generally higher in hippos than in zebras supporting the contention that the pseudo-ruminant condition is advantageous.

Table 2.9 The digestion of food components by *Hippopotamus amphibius*. (After Abaturov *et al.* 1995.)

Item	Mean % Food	Dry Matter in Faeces %	Digestibility Coefficient
Crude Protein	9.61	5.74	67
Lipids	1.06	1.16	39
Crude Fibre	30.15	23.59	57
Lignin	16.33	29.43	0
Nitrogen Free Extract	34.45	27.07	56
Ash	8.38	13.01	14

ENERGETICS OF HIPPOS

The common hippo, and probably the pygmy, leads a lazy life style that is very economical with energy. It spends the bulk of its time resting in the water and doing very little, unless it is a male that needs occasionally to rebuff a challenge to its territorial status. It may drag itself onto land to bask in the sun but such activity is hardly debilitating. Immersion in the water saves it from heat stress and the buoyancy lightens its body and takes the weight off its feet. Admittedly it has to walk to its feeding grounds, sometimes over distances of several kilometres, but it does so at night when there is little thermal stress. This conservation of energy may be the reason why the hippo takes in a quantity of food that is well below that of mammals of similar size. Some details are given later in the chapter on feeding behaviour, where it will be seen that the average weight of the stomach contents in a sample of Ugandan hippos represented some 0.95% of the body weight in males and 1.3% in females. This is only about half of the equivalent figures for other large mammals, e.g. 1.8 to 2.4% for elephants from various national parks in Uganda (Malpas, 1978).

Wright (1987) analysed the energetics of a tame hippo resting on a concrete floor. In such an animal, about 75% of the surface area is irradiated by the sun, a figure similar to that for a person standing upright. Hence the mean radiant temperatures calculated for man can be used in the estimation of the heat load experienced by the hippo through radiation and convection. If the hippo is considered as a black body with an absorbance of 1, it is possible to estimate the radiant heat flux down the gradient from radiant temperature (Tr) to skin temperature (Ts). Under cloudy conditions Ts usually exceeds ambient temperature (Ta) so that the hippo can lose heat to the atmosphere and is able, therefore, to leave the water and lie on the shore. If the sun is shining, Tr soon exceeds Ts and the hippo is obliged to lose heat by wallowing or "sweating".

Fig. 2.12 illustrates the fluctuations in environmental and bodily measurements taken on a sunny day with the tame hippo. It can be seen that skin temperature (Ts) increased initially with Tr, which minimised heat gain from the environment. This gain was 217 W m^{-2} at the highest radiant temperature around midday but this would have been 23% higher if the skin temperature had not also risen. During the afternoon increased evaporation reduced the skin temperature with the effect of slightly increasing the radiation transfer to a maximum value of 228 W m^{-2} at 15.00h. Evaporation from the skin is induced by an increase in the vapour pressure at the skin surface, which in turn results from an increase in Ts with Tr. As Ts fell as a result of this evaporation, the vapour pressure difference between skin and surroundings (Ps × Pa) also fell and was probably responsible for much of the unevaporated secretion shown on the figure.

Despite the fluctuations in skin and radiant temperatures, the core temperature of the hippo remained remarkably constant to within one Celsius degree throughout the day. The metabolic heat production of this particular hippo of 860 kg was 2575 kJ kg^{-1} h^{-1} as measured from oxygen consumption on land. An increase of one degree in core temperature represents a heat storage of 3010 kJ over the eight-hour observation period or about 17% of the day-time metabolic heat production.

These experiments imposed much harsher conditions on the hippo than it would have experienced in the wild as it was denied access to its pool. Under natural conditions the animal would have moved into water as a means of cooling off but the results do show that a hippo is capable of maintaining a constant body core temperature even while on land.

The conclusions from these experiments on a tame hippo seem to apply to wild animals. The basic metabolic rate (BMR) of the common hippo was estimated by Wright (1964) by applying the mouse–elephant scale (Benedict, 1936) to measurements of the skin area relative to body weight. This gave a value of 61 kcal. m^{-2} h^{-1} for an animal weighing 1500 kg. In order to calculate the total heat production of a hippo the heat from digestion must be added. This is about 50% of the BMR in cattle and, applying the same rate to hippos, a total figure of about 90 kcal. m^{-2} h^{-1} is obtained. This heat must pass from a wallowing hippo into the water through its skin. Calculations of the thermal conductivity of the skin suggest that there is a temperature gradient through the skin's thickness of 1.9 to 2.8°C cm^{-1}. Actual measurements showed a gradient of 1.7 to 2.4°C cm^{-1}, suggesting the estimated BMR is reasonably accurate.

Wright constructed a table showing the thermal balance in two wild hippos shot in Uganda, one of which (Hippo A) was taken in late morning while lying in the shade and the other (Hippo B) early in the night, when there could have been little energy expenditure due to digestion although some due to walking. His table is

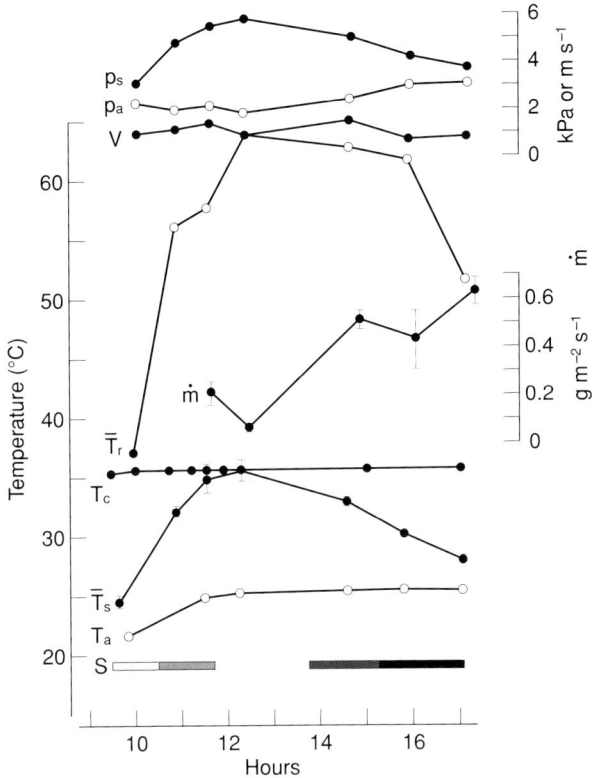

Figure 2.12 Body core temperature of a tame common hippopotamus in relation to environmental factors on a sunny day in Kampala, Uganda. The hippo was denied access to its pool. Note that the core temperature remained constant throughout, despite large changes in other factors. Key: P_s – saturated water vapour pressure at mean skin temperature; P_a – ambient water vapour pressure; V – wind speed; T_c – Body core temperature; T_s – Mean skin temperature with standard errors; T_a – ambient temperature; T_r – mean radiant temperature; m. – evaporative water loss with standard deviations; S – visual assessment of general skin appearance from (left to right) very light, light, free and copious "sweat". From Wright (1987).

reproduced here as Table 2.10. The metabolic increments for standing, walking and digestion are taken from measurements on cattle and horses and may not apply to hippos, whose digestive efficiency appears to be less good than that of some other ungulates.

The hippo resting by day (Hippo A) loses and gains less energy than the one grazing at night. The objective of the estimates, however, was to consider the effects of a rising air temperature on a hippo if it remained on land all day. This would reduce the heat loss due to radiation and could even reverse it. It would also reduce heat loss due to convection, unless the wind velocity also increased. Even if it did, the effect would be minimal as the loss varies directly with the skin/air temperature difference but only as the square root of the wind velocity.

Table 2.10   Thermal balance factors in two wild hippos from Queen Elizabeth National Park, Uganda. (After Wright, 1964.)

	Hippo A	*Hippo B*
Energy loss (kcal. h^{-1})		
Radiation	298	363
Convection	261	285
Radiation	176	155
Insensible water loss	120	100
Total Loss	**855**	**903**
Energy gain (kcal. h^{-1})		
Basal Metabolic Rate	612	612
Standing (+ 10%)	0	61
Walking (+ 37%)	0	249
Digestion (+ 50%)	306	0
Total Gain	**918**	**922**
Environmental factors		
Air temperature	29.7°C	23.1°C
Relative humidity	44%	80%
Wind velocity	15 m/min	34 m/min

This leaves only "sweating" as a means of shedding excess heat. As has been shown, sweating through the evaporation of red skin secretion may be sufficient to control body temperature although that is not the primary function of the skin glands. A free-ranging hippo, however, is not dependent on sweating as it can cool itself by moving into the water. This is a second reason for the amphibious habit of hippos, the first being to prevent the skin from cracking by drying out. What is not clear, of course, is whether the hippo became aquatic because of its skin anatomy or whether its skin anatomy evolved because of its aquatic habits.

CHAPTER 3

THE ORIGINS OF HIPPOS

The extinct Hippopotamus gorgops

HIPPO ANCESTRY

Hippos are classified as members of the Suiformes, a sub-order of the Order Artiodactyla, the even-toed ungulates. Most artiodactyls have two functional toes and are often called cloven-hoofed as a consequence but the hippos have four toes, all of which touch the ground. The Suiformes is a large group containing 12 families of which all but three are extinct. The living forms are the true pigs (Suidae), the (mainly) South American peccaries (Dicotylidae) and the hippos (Hippopotamidae). The relationships of the hippos to other families of the Suiformes are obscure but of the extinct groups, the family Anthracotheridae is thought by some palaeontologists to be a likely ancestral form. The taxonomy of hippos is not generally agreed but many systematists group hippos with anthra-cotherids in the super-family Anthracotheroidea. The anthracotherids were large, pig-like animals that were probably semi-aquatic. Their similarities to hippos rest in the dentition but their remains have rarely been found with those of hippos, which does not support their supposed ancestral status. If they were indeed ancestral, the hippos split off from them during the Upper Oligocene to Lower Miocene, i.e. about 25 million years ago.

The hippos are now confined to Africa but in prehistoric times they had a much wider distribution. The first representatives of the group are known as fossils from the Middle to Late Miocene of East Africa so it is likely that, as with so many other animal groups, hippos originated on that continent. The first hippopotamids from outside Africa appear in Eurasia during the Late Miocene. By the Pleistocene, a million or more years ago, hippos had spread throughout Europe and Asia but they never reached the New World or Australia. They were present in Britain up to about 120,000 years ago (Gascoyne *et al.*, 1981).

MODERN HIPPOS

The division between the two modern species of hippo seems to be of long standing as most of the fossil forms since the late Miocene can be assigned to one or other lineage. It is not easy to distinguish the two genera from fossil evidence although the difference is distinct if sufficient material is available. Several anatomical features, mostly dental, separate the two. Size is not a good factor as there are large species of *Hexaprotodon* and small species of *Hippopotamus*. *Hexaprotodon* tends to be more "gracile", i.e. more delicate, even when large. This is indicated by the ratio between the length and width of the limb bones. The name "hexaprotodon" refers to the three pairs of incisors found in some fossil forms but the modern *Hexaprotodon* has only one pair or, rarely, two. The large hippo has two pairs, i.e. it is tetraprotodont, with the central pair being much the larger. Specific differences in the insertion and occlusion of the incisors were mentioned in Chapter 2. A further difference is the deeper groove on the back of the upper canine in *Hexaprotodon* and the double-rooted first premolar in the upper jaw. There are also cranial differences particularly in the nasal, frontal and lacrimal bones.

FOSSIL HIPPOS

The discussion of hippo ancestry becomes a little confusing if one conforms to the convention of abbreviating the generic name to a single letter (as in *H. amphibius*) as it does not indicate which genus of hippo is being considered. Therefore, in this chapter I shall follow the excellent, if unconventional, example of some hippo palaeontologists and refer to *Hippopotamus* as *Hip.* and *Hexaprotodon* as *Hex.*

There are no fossils that can be identified with the modern pygmy hippo although there are, of course, fossil species of *Hexaprotodon*. Two dwarf hippos were recently described from sub-fossils found in Madagascar by Stuenes (1989). One of these, *Hip. lemerlei*, showed skull features that allied it to the living common hippo. There were consistent size differences in the skulls, suggesting that this was a sexually dimorphic species. The other species was also assigned by Stuenes to the genus *Hippopotamus* as *Hip. madagascariensis* although in its cranial anatomy, it showed similarities to

Hexaprotodon and would be placed in that genus by most taxonomists. Nevertheless, Stuenes assumed that the two were closely related and that both should be considered as belonging to the same genus, *Hippopotamus*. There is a suspicion that he took this view on grounds of incredulity over the possibility that two hippos could have succeeded in reaching Madagascar from Africa but he could be correct over the taxonomy as parallel evolution in island species of *Hippopotamus* is known. The common feature is an adaptation to a cursorial life, i.e. to running.

There is no convincing evidence one way or the other for the origin of the Madagascar hippos but it is best to take a conservative view and assume, until it can be proved otherwise, that if there were two genera, they arrived separately. The situation is complicated by the subsequent discovery in Madagascar of a third extinct species of hippo, *Hip. laloumena* (Faure & Guerin, 1990). This was a large animal, although not as big as the present-day common hippo, and quite clearly belonged to the genus *Hippopotamus*.

Hip. lemerlei became extinct only relatively recently – estimates vary but it was alive at least until 1000 AD give or take a couple of hundred years. It was probably killed off by man following the human settlement of the island some 1500 to 2000 years ago. The other two species have not been dated but they are believed to have existed in Holocene times, i.e. since the Pleistocene, possibly within the last few thousand years.

The co-existence of three species of hippopotamus in Madagascar makes it highly likely that the island was invaded three times by hippos from the mainland of Africa, once by *Hexaprotodon* and twice by *Hippopotamus*. Although the Mozambique Channel is only some 400 km across at its narrowest point, it has proved to be an insuperable barrier for many would-be colonisers. Madagascar split away from Africa about 150 million years ago, i.e. during the time of the Jurassic dinosaurs, well before mammals evolved into the wide variety of orders and families that we see today. This is why the Malagasy fauna and flora are so peculiar, with no carnivores other than viverrids and no primates other than lemurs. Almost all groups of plants and animals are highly endemic with high percentages of species that are found nowhere else. It is unlikely that many mammals, except bats, reached Madagascar after the split and those that did, found themselves without competitors and spread into niches not typical of their group, e.g. lemurs show convergent evolution with squirrels, mice and monkeys. Ungulates never made it except for the hippos, which would have had fewer difficulties than most in crossing the sea and it is not, therefore, inconceivable that there were three successful colonisations.

A fossil dwarf hippo belonging to neither modern genus is known from the Pleistocene of Cyprus. This is *Phanourios minor*, which shows similarities to both although on balance it seems closer to *Hippopotamus*. Coryndon (1977) places it in that genus on the basis of its skull anatomy but Houtekamer & Sondaar (1979) strongly disagree. An examination of the forelimb suggests a more cursorial (fast running) life-style than that of the modern forms, e.g. the shoulder blade was ruminant-like, the leg bones were more slender and the toes were much shorter. The limb proportions were very different and the cushion or pad found on the foot of living forms could not have existed in *Phanourios*. Houtekamer & Sondaar believe that the animal occupied a different niche, with a foot that was adapted for running over the stony ground that is still a feature of Cyprus today.

Fossil Hexaprotodon

Fossil species of *Hexaprotodon* are known from other regions of Africa. They include *Hex. imagunculus*, which was found in the western Rift Valley of East Africa in deposits dating from the late Pliocene to early Pleistocene, and *Hex. aethiopicus* of Kenya, which was recovered from the Turkana Basin in the main Rift Valley from rocks of the same geological age. These two species used to be assigned to the *Hippopotamus* genus. A small fossil hippo from fairly recent deposits in Algeria, *Hex. hippoensis*, may be closer to *Hippopotamus* than to *Hexaprotodon*.

There are several large *Hexaprotodon* fossil species such as *Hex. harvardi* from Lothagam and *Hex. sahabiensis* from Libya (Gaziry, 1987). Fossils of three large hexaprotodonts have been described from the Turkana Basin, namely *Hex. protamphibius, Hex. karumensis* and *Hex. shungurensis*. These were alive at the same geological time and may have co-existed. Two *Hexaprotodon* species were described from Hadar, one has not yet been named but the other has been described as *Hex. coryndonae*. A third species was found at the same site and assigned to the new genus *Trilobophorus* on account of differences in the lacrimal region of its skull. Further fossils are known from East Africa but they have not so far been described.

Hexaprotodon species appear to have been the dominant hippos in East Africa during the late Pliocene and early Pleistocene. The specific validity of some of the remains is less than certain and Harris (1991) raises the possibility that *Hex. shungurensis* was not a separate species at all but was the female of the larger *Hex. protamphibius*. As mentioned earlier, one of the Madagascar hippos, *Hip. lemerlei*, was probably sexually dimorphic and it may be that some of the hexaprotodont species were as well. On the other hand, *Hex. shungurensis* could be a genuine species and be ancestral to the smaller *Hex. aethiopicus* as it is intermediate in size between it and *Hex. protamphibius*.

Fossil Hippopotamus

The fossil history of the *Hippopotamus* genus is poor and, compared with many other groups of large mammal, it is sketchy in the extreme. The oldest fossil hippo in the genus was *Hip. kaisensis* from the western Rift Valley of East Africa. This was very similar to the modern hippo and could be its ancestor if not identical to it. Other *Hippopotamus* specimens have been found elsewhere in East Africa. A distinct species was *Hip. gorgops*, distinguished by its tall eye sockets. This was a large hippo, larger even than *Hip. amphibius*, and it occurred in East and southern Africa but not outside the continent except possibly in Israel, where dubious fossil remains were located. The hippos found in Europe and Asia are referred to as *Hip. major* or *Hip. incognitus*. These were very similar to *Hip. amphibius* and could be no more than subspecies. Dwarf species of *Hippopotamus* are known from Mediterranean islands such as Malta and Cyprus as well as from Madagascar. There is no significance in this as island dwarfing is a common phenomenon in large mammals. Nevertheless, the remains of an undescribed dwarf *Hippopotamus* have been found on the African mainland at Olduvai Gorge in Tanzania. Dwarfing is a relative term and some so-called dwarf species were not all that small.

HIPPOS IN BRITAIN

Fossil remains of hippos are quite common in Britain although they are confined to England and Wales with a preponderance in East Anglia (Stuart, 1986). A map showing the distribution of fossil hippo material is reproduced as Fig. 3.1. A lack of fossils does not necessarily mean a lack of hippos and it is very likely that most of the rivers in southern Britain were swarming with hippos, much like the African rivers today. These fossils date from the Upper (late) Pleistocene and belong to two species, the existing *Hip. amphibius* and an extinct form *Hip. major*. Both species were also found in Europe. *Hip. major* was not very different from the present-day hippo but it had a much longer snout. It also had a differently shaped lower jaw, whose undersurface was concave when viewed from the side, unlike the convex shape of the living hippo, although this distinction is not always clear in some remains.

The British fossils appear to have lived in one or other of two interglacial periods, the Cromerian and Ipswichian, which occurred about 350 000 and 120 000 years ago

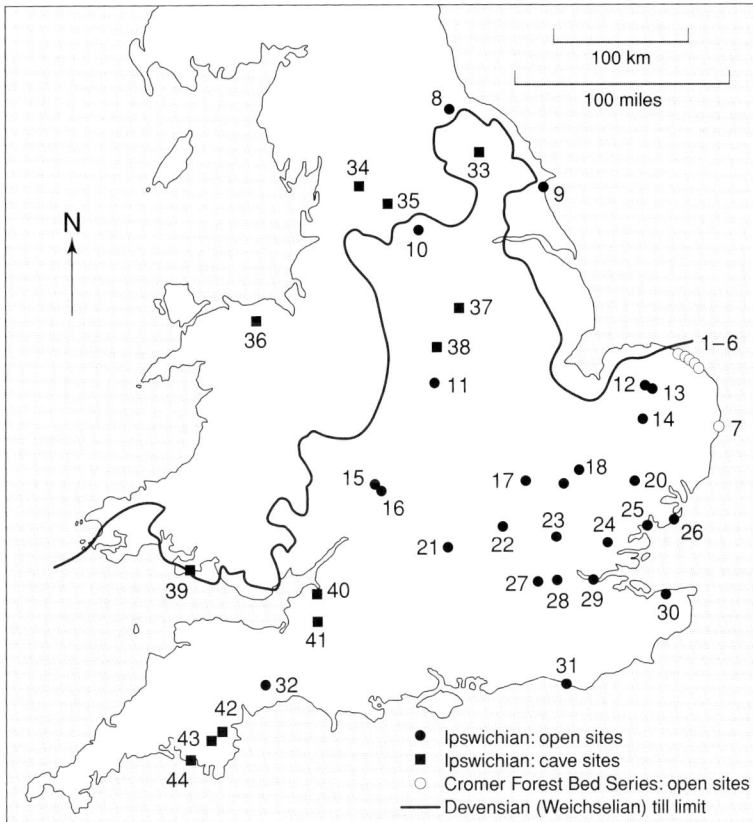

Figure 3.1 Sites where fossil remains of *Hippopotamus* have been found in Britain. From Stuart (1986).

respectively and which are named from the towns nearest to where the first fossils were found, namely Cromer and Ipswich. The Ipswichian was the last interglacial to occur and the hippo remains were associated with those of many other large mammals, suggesting that the hippos were part of a wide-ranging and diverse community. This is hardly surprising for the Pleistocene fauna of large mammals would stand comparison with that in the most diverse African savannas of today. It can be deduced from the pollen grains found with the fossils that during both interglacials the vegetation was typical of a temperate climate, probably not very different from that today.

All the remains from the Ipswichian beds can be assigned to *Hip. amphibius* but the earlier Cromerian fossils are more ambiguous. Most of these are clearly *Hip. amphibius* but the only find that was definitely *Hip. major* could not be dated as its sediment matrix did not contain pollen and it could possibly be from an earlier interglacial period.

A feature of the Pleistocene hippo fauna throughout the world is the number of species that were present at the same time. This contrasts markedly with the situation today with just one large and one small representative. It would be fascinating to know if these extinct species actually co-existed, i.e. if they shared the same area at the same time, and if so, how they avoided competition on the one hand and interbreeding on the other. Assuming that they had similar feeding habits to the modern species, it is difficult to see how more than one grass-eating hippo could occur in the same area.

EXTINCTION OF HIPPOS

On a palaeontological scale, extinct hippos lived until quite recently and some of them certainly co-existed with man as hippo remains have been found on Cyprus and Madagascar at sites, no older than 2000 BP, showing evidence of human habitation. Some remains associated with people are much older. Simmons (1998) found the remains of *Phanourios minutus*, a pygmy form, on Akrotiri Peninsula south of Limossal in Cyprus in what appears to have been a midden. Some of the hippo bones had been cut or broken open and some had been burnt, suggesting that the meat had been cooked. This raises the possibility that their extinction was caused by early human hunters. The most likely alternative to the human factor in the decline of *Phanourios* is climate change but the evidence for any profound change is ambiguous. Incidentally, radio carbon dating of surrounding material puts the date of the site at around 10 200 years BP so extending the period of human settlement on the island by 1200 years from the previously accepted figure of 9000 years.

The fact that people ate hippos does not necessarily mean that humans caused their extinction but this would have been easier to bring about on islands than on the mainland because hippo populations would have been smaller and unable to escape by emigration. This points the finger at man, and human hunting was probably the cause of island extinctions in the prehistoric times, but it does seem improbable that the relatively few people who were abroad at the time could have wreaked such havoc on mainland populations.

Not only hippos became extinct at the end of the Pleistocene; many large mammal species also succumbed, and no satisfactory explanation for this mass extinction has been put forward. There is no evidence of a collision with a meteorite or comet similar to that which appears to have put paid to the dinosaurs. Possibly no one factor was responsible and it may be that climate change started the slide towards oblivion which was exacerbated by human hunting and human alteration of the environment through the spread of pastoralism and, later on, of agriculture. An awkward problem in any hypothesis is Africa, which seems to have escaped mass extinctions despite being the cradle of the human race.

HIPPOS AND WHALES

Hippos share with whales a tendency to be aquatic. There is no reason to suppose that this is other than a coincidence but it is intriguing to learn from recent research that, genetically, the two groups show similarities and that there is evidence of a common ancestry. The fossil history of the cetaceans (whales, dolphins and porpoises) is not much better than that of hippos and there is no agreement over their origins. Perhaps the generally accepted opinion is that cetaceans are descended from early ungulates that became aquatic, rather like hippos, but an alternative view is that they are derived from primitive carnivores. Evidence for a hippo/whale link was found through the sequencing of the mitochondrial DNA cytochrome *b* gene by Irwin & Arnason (1994), who suggested for the first time that hippos and whales share an evolutionary lineage. Support for this idea came from an examination of milk casein genes by Gatesby *et al.* (1996). Their analysis of kappa-casein (Exon 4) and beta-casein (Exon 7) protein genes in milk taken from cetaceans and other placental mammals suggested that hippos were more closely related to cetaceans than to any other artiodactyl. An analysis of the nuclear casein sequences combined with published mitochondrial DNA data also supported the hippo/cetacean link.

Hasegawa & Adachi (1996) are more cautious in their conclusions from an analysis of mitochondrial DNA sequences. Although this showed the ruminant/cetacean relationship, it did not suggest the existence of a ruminant/suiformes clade although the possibility cannot be ruled out. In other words, the traditional assumption that whales evolved from the ungulates is supported but not the assumption that pigs, peccaries and hippos are closely related to the other artiodactyls. It follows, therefore, that the link between hippos and whales has not been demonstrated. These authors also examined other published data on cytochrome and haemoglobins and concluded that the evidence for a hippo/cetacean link is "fragile".

Randi *et al.* (1996) sequenced the mitochondrial cytochrome *b* from four species of suids and compared their results with those obtained for the common hippo by Irwin & Arnason (1994). Their results confirmed the wide genetic separation between hippos, peccaries and pigs but did not help in establishing the phylogenetic position of the hippo in relation to cetaceans.

A thread running throughout these genetic analyses is one supporting a weak but persistent link between hippos and whales, suggesting that both share a common ancestor. If this is true, then the shared aquatic habits may not be coincidental and hippos may, after all, be the missing link between whales and terrestrial ungulates. This is not to say that hippos evolved into whales but the ecology of the present-day hippo could reflect that of those artiodactyls that first returned to the sea and became cetaceans. The paucity of the fossil record for both groups allows such speculation, for the presumed split between hippos and whales occurred long before the first hippo bones became fossilised. More research is obviously required, both on the palaeontology and on the genetics of the two groups, before any firm conclusions can be drawn.

CHAPTER 4

THE SOCIAL LIFE OF HIPPOS

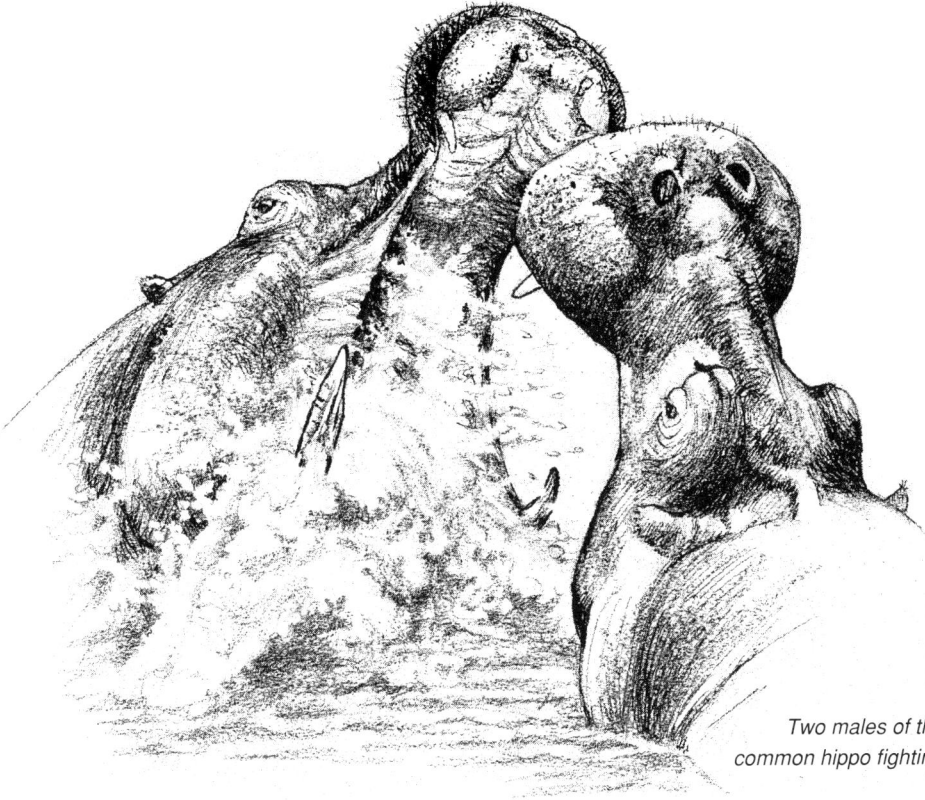

*Two males of the
common hippo fighting*

Not a lot is known about the social life of hippos and much of what has been written about this aspect of their biology is incorrect, particularly the details of territorial behaviour in male *H. amphibius*. The reason for this lack of information is not difficult to find. Most information on the behaviour of large mammals is derived from studying the behaviour of marked or individually known animals. Most progress is made if it is possible to track specimens to which radio transmitters have been attached. This is particularly useful for locating animals in thick country.

The hippo seems to have been designed to frustrate the ethologist who may wish to mark one for study. This involves first catching the animal, usually through remote delivery of a tranquillising drug, but the practice has proved to be extraordinarily difficult with hippos. In the first place, their reaction to drugs is unpredictable. Sometimes the drugs work well and the animal collapses but more often there is little or no reaction. It may be that the very thick skin prevents the drug from being properly injected. Even if the drug works, there may be problems, for the hippo's first

47

reaction is to head for water, and if it makes it before collapsing, it invariably drowns. Consequently, it should be darted as far away as possible from water but the only time this situation obtains is at night, when it is difficult to follow a darted animal that is running off at top speed. Sometimes the hippo is out grazing by day but at such times it is usually fairly close to water. On one occasion when I was present, a hippo was immobilised by day but only the hindquarters of the animal had been affected, leaving the head free to move and snap at the scientists. As the object of the exercise was to collect respiratory gases, this was rather awkward. This is unusual, however, but even a completely sedated hippo presents difficulties.

Let us suppose that all has gone well and that a hippo has been drugged in the open by day. The problems are far from over. Large mammals are usually marked with ear tags or a collar. Sometimes a number is painted on the skin. Painting a hippo is not very rewarding, even with quick-drying, waterproof paint, since not much of the body is visible under the usually murky water. Ear tags tend to be quickly lost since the first thing a hippo does on surfacing is to waggle its ears vigorously with the result that metal tags slice through the ear cartilage and are lost. Plastic tags are little better in this respect as well as being more fragile. If a radio transmitter is to be fitted, where can one attach it? The normal place is on a collar around the neck but a hippo does not have much of a neck and a collar, if it is not to throttle the animal, tends to slip off over the head. The transmitter, of course, has to be particularly water repellent.

The handling of an animal can be avoided if it can be identified from natural markings as, for example, the stripes on a zebra or tiger, which are unique to the individual. Unfortunately, one hippo looks rather like another but they can be singled out through scars, particularly those on the bodies of territorial bulls. The problem here is that the body is obscured in the water and, on land, the scars do not show up very clearly at night. There is the further problem that the scars are being continually added to in sexually active males. Nevertheless, the use of natural markings has proved to be successful and is probably the best technique to use for studying the behaviour of the species.

These difficulties refer to the large hippo but most also apply to the pygmy species, which presents other problems of its own. It is, for example, very rare so that it is unusual even to see an individual, let alone catch one. It too is nocturnal but it lives in thick forest, not the open grasslands of its larger cousin, making it very difficult to follow for any distance. The best way of catching the pygmy hippo is probably to trap it as it is used to entering small spaces such as holes and tunnels.

THE BEHAVIOUR OF *HIPPOPOTAMUS AMPHIBIUS*

One person who has succeeded in surmounting the problems of catching hippos is Hans Klingel (1991), who was able to recognise over 200 hippos in the Queen Elizabeth National Park, Uganda, from various natural marks such as misshapen or torn ears, broken tails and even variations in colour patterns. He also marked 20 animals at night by spraying them with paint from a moving vehicle. This proved more useful than natural marks in the dark but the paint lasted only for about three weeks before fading. He had problems with drug immobilisation and abandoned the technique after trying a variety of drugs such as etorphine, fentanyl, azethylpromazine and sernylan, all of which proved to be unsatisfactory. Nevertheless, seven animals were successfully captured at night and fitted with ear tags and streamers as well as being painted.

Klingel (1991) studied the animals in two areas of the park; one was 1200 m of the shore of Mweya Peninsula, which juts into a large lake (Lake Edward) and the other was 600 m of the Ishasha River in the south of the park. The following account leans heavily on Klingel's work for it is the only good study available. Although the observations were restricted to a small part of a single national park, the conclusions are likely to be generally applicable. Klingel made less intensive observations in other habitats such as the Mara River, Kenya, the Okavango Delta, Botswana, and the Luangwa and Kafue Rivers in Zambia, and could find no essential differences in behaviour. Hence it is likely that his Uganda observations are typical of the species.

Social life

Social life for hippos tends to be confined to the river or whatever body of water in which they spend the day. Hippos are territorial only in the water, with the males holding a linear territory consisting of the shoreline and a narrow strip of the bank. This is defended against other bulls but only if they challenge the incumbent male, which is otherwise tolerant of them provided they behave submissively towards him should they happen to meet. Consequently, the territorial system is not at all obvious, with adult bulls spread along the shore line, although many of the younger males tend to gravitate towards the boundaries between territories. This general mixing of animals of all ages and both sexes has led some authors to deny the existence of territoriality (e.g. Olivier & Laurie, 1974a) and it was only when individually recognisable animals were studied over an extended period that territoriality was identified.

The social significance of hippo groupings was for long obscure and the use of the term "school" rather than "herd" reflected that uncertainty. It seems, however, that rather than there being a random mixture of animals, there are three basic groupings of hippos in a school. First, there are the territorial males, which represent about 10% of the population and which are technically solitary although mingling with the other animals. Those males that are not territorial form bachelor groups, often within a territory. Finally, there are the female groups. On the whole the sexes remain segregated although a few members of the opposite sex are often found within a male or female gathering. There are no social bonds between the adults within a group despite the fact that hippos seem to love to lie in close contact with each other.

Territoriality

The sizes of the linear territories found by Klingel (1991) in Uganda varied considerably. They measured from 250 to 500 m on the lake shore but they were much smaller in the Ishasha River, where they were only 50 to 100 m in length. The river territories included both banks but there was no defined outer limit for the lake shore territories. Where hippos spent the day in swamps, the territories followed a mosaic pattern.

Hippos probably hold their territories for life or for as long as they can manage to do so. As with many ungulates, territorial males assert their status by adopting a "proud" stance with the head held high and ears cocked. Their ownership may be challenged from time to time by a hopeful bachelor and it is then up to the holder to intimidate his opponent by threats or in the last resort by physical attack. Conflicts between two territorial males at their boundaries evokes only a ritual fight, which amounts to little more than splashing water over each other and defaecating simultaneously while standing side by side but if a bachelor male attempts to supplant the territory holder, or if a territorial male tries to enlarge his territory, the consequences are much more violent. The pair again stand side by side, nose to tail, but instead of splashing each other, they deliver vicious swings of the head, gashing the opponent's flank with the huge canine teeth. Although the hide is 5 cm (2 in) thick in this region, serious wounds may be caused that sometimes penetrate the peritoneum and result in death. For the most part, the wounds heal up very quickly, lending support to the belief that the "red sweat" has antiseptic properties.

Fights of such seriousness are rare and the general relationship between hippos in the water is one of friendliness. There is no obvious hierarchy and no one hippo is dominant over another, except for the territorial males.

The females live within a territory, to which they return each morning after a night's grazing. Each territory, therefore, has its "school" of females but there are no social bonds between the individuals, except for mothers and daughters. The association between females depends solely on their choice of which territory in which to reside. The territory holder has exclusive mating rights to these females and it is obvious that the value of the territory depends on the number of females that it attracts. Female choice may be based, in part, on the quality of the male but it is more likely to depend on topographical features such as the depth of water, lack of a strong current and the presence of gently shelving sand banks on which to loaf. The physical quality of the territory may change, e.g. if the river should alter its course. Klingel recorded several such changes, particularly on the Ishasha River, where some good territories changed to poor ones and *vice versa*. One branch of the river dried up and the territorial bulls were obliged to leave and seek other territories. Some entered a newly formed swamp and carved out fresh territories for themselves.

The number of females in a male's territory has probably never been measured accurately. Klingel maintains that the number fluctuates, with some hippos frequently leaving the group and others joining so that the number present may vary from as little as two to more than one hundred. Laws & Clough (1966) reported the average size of a school to be ten with a maximum of 107 animals. It is difficult to be precise for it is not easy to count hippos in the water as they are not all visible at the same time.

The male's hold on its territory depends on its strength but this is not the whole story for as is clear from the previous paragraph, chance can play an important part. A bull whose territory has dried up has no choice but to look for another or to give up and join a bachelor group. Its chances of success will depend on its fighting ability but also on the number of potential rivals. When or if the channel floods again, the original owner, if it wished to reclaim its territory, would have to contend with bachelors that may have seized the chance to stake out a territory for themselves.

Competition between males appeared to be more intense in the Ishasha River than in the Mweya study area judging from the instability of the habitat and the density of the hippos. In the river, there were 33 hippos for every 100 m of shoreline compared with only seven for Mweya Peninsula. Hence there were many more hippos competing for resources and because of the changing nature of the habitat, many more opportunities for competition. These differences were reflected in the length of tenure of the territories.

The duration of tenure on Mweya Peninsula sometimes extended for the whole period of observation (four and a half years). This was true of bulls on four of the six territories studied and two of those bulls were still in residence when Klingel made a return visit almost eight years after the start of the study. One of them was still there after 12 years. One cannot be certain that it held the territory continuously but it is very likely that it did so. This contrasts with the situation in the Ishasha River, where the recorded tenure could be as little as a few months and never exceeded two years.

The territories of hippos are established to defend mating rights and not food. Hippos are not territorial away from water despite statements to the contrary in the literature, e.g. by Hediger (1951) who believed that males defended pear-shaped territories on land. Although the quality of a territory may deteriorate from causes over which the holder has no control, it pays a hippo to stick to its territory rather that to try its luck elsewhere, for the costs of failure are severe. It is certainly better to hold on to a deteriorating territory rather than to abandon it and join the ranks of the bachelors. Territories can improve as well as decline and before long, the hippo may find itself in charge of a superior territory once again.

The females are in no sense territorial and are not necessarily confined to a single territory although, as mentioned above, most return to the same one after grazing. It is not unknown, however, for females to use more than one region, e.g. a few animals alternated between the northern and southern shores of Mweya Peninsula in Klingel's study. The day-time home range of females is, however, about the same size as a male's territory, i.e. up to about 200 m of shoreline.

Defaecation

Defaecation plays a prominent role in the life of a hippo and often involves more than the mere elimination of waste. The hippo's tail is a very efficient muck spreader and is wagged vigorously during defaecation so that the dung, which is loosely constituted, is widely scattered. Very often the action is performed with the rear pressed up against a bush. As the dung piles are spread around the grazing areas, it used to be thought that such bushes are territorial markers but it is now well established that the hippo is not territorial on land. It is tempting to think that the dung-sprayed bushes must have some function and it is possible that they are used as beacons to help the hippos navigate around their home range at night. This has been proposed by Boulière & Verschuren (1960) and Curry-Lindahl (1961). It is true that the dung piles are usually on or near the hippo trails and they could well serve as orientation clues although they would seem unnecessary as the hippo has only to keep to the trail to find its way back to the water. It may be that the dung piles act as message points to give information about the identity of individuals in the area. Urine is often voided at the same time as faeces and it could add to the value of the deposit as a signal, for secretions from scent glands are passed out with the urine in many species.

Hippos also defaecate over each other in a manner thought by Verheyen (1954) and Curry-Lindahl (1961) to have social significance. This idea is supported by observations made on the Mara River in Kenya by Olivier & Laurie (1974a) who, contrary perhaps to expectations, saw subordinate animals turn around and spray dominant bulls in the face with their faeces. Even if no dung was produced, water was almost always splashed onto the face of the dominant animal. They never saw such behaviour between two dominant males. Dung spraying by large males in the water may also take place without the involvement of another animal. More usually, however, the male leaves the water, or moves over to an exposed rock, and defaecates onto an established dung pile. Young animals often follow the bull and take a great interest in the faeces, even to the extent of eating the dung. Adult females will also sniff the dung but Olivier & Laurie say they never saw one eating it.

This habit of wallowing hippos of walking over to a dunging site in order to defaecate is confined to the male and is considered by Olivier & Laurie to be sufficient to identify it as a male. Verheyen (1954) goes further and believes that it is only males that mark the inland bushes.

Observations by Klingel (1991) suggest that the tables may sometimes be turned, with the dominant male spraying the subordinate. A bachelor hippo seems to go out of its way to placate the dominant bull. On land it does so by approaching from a distance and dropping to a crawl with lowered head when within a few metres. It then sniffs the genital region of the territorial male, who completely ignores it although he is almost certainly well aware of its presence. Sometimes the dominant animal defaecates over the subordinate but it is not clear whether this is incidental or whether it is a sign of appreciation of the deference shown. The appeasement ritual is similar in the water except that the two may meet nose to nose.

What all these observations suggest is that the faeces have a social function. One cannot separate faeces from urine in this respect as hippos may direct a spray of urine over the dung as it is voided. Hippos certainly seem to take an interest in each other's rears by smelling or licking the anal region and by following each other nose to tail both in the water and on land. Olivier & Laurie (1974a) recorded this behaviour on 21 occasions and found that it occurred between most age and sex classes. An adult male was followed by another adult male only once, by a subadult three times and by a juvenile seven times. A subadult was also seen following another subadult three times and an adult female once. An adult male following an adult female, which has a more obviously sexual connotation, was observed six times. It is very likely that hippos can recognise each other as individuals from scented compounds in the faeces or urine or both. In the case of a female, it is probable that its reproductive status is also identifiable by smell.

Grazing routine

When grazing at night, all hippos do so as individuals whether they be cows, bachelors or territorial bulls. The only exception is that young calves and subadults accompany their mothers, remaining with them until almost full grown, i.e. between six and eight years of age. Consequently, a female may have several calves of various sizes in tow. When moving through the bush, she takes the lead, followed by the calves in ascending order of age. Females do not take newborn calves ashore but remain in the water, apparently without feeding, for several days. Young hippos are vulnerable to predation particularly from lions and although the mother is a formidable protector, she may not always be quick enough to foil an attack, which could prove fatal for very young babies.

The grazing grounds are some distance from water and the hippos have to walk several kilometres to reach them. Some estimates have been made of the distances involved. Lock (1972) put the figure for the round trip at 3.2 km in the Queen Elizabeth National Park in Uganda. Olivier & Laurie (1974a) investigated the grazing range of hippos from the Mara River in the Serengeti National Park by following hippo trails. They found that hippos grazed within 20 m of the end of a path and on that basis they calculated the grazing range to extend for an average distance from the water of 1.35 km for the northern bank and 1.05 km for the southern. These are rather shorter than the figure of 2.2 km, with an average distance of 3.5 km for the round trip, given for the Chipinda area of the Gonarezhou National Park in Zimbabwe by C.S. Mackie (O'Connor & Campbell, 1986). O'Connor & Campbell provide more detailed figures for a traditional hippo wallow at Nyahungwe, also in the Gonarezhou Park (Table 4.1). These show that there is a significant difference between the seasons, with the hippos travelling farther from water and covering greater distances in the dry season than in the wet. This might be expected from the poorer grazing available in the dry season.

Hippos find their grazing grounds by following well-worn trails, which may be so eroded by generations of hippo feet that they become almost like trenches. These are the game trails that are such a notable feature of African savannas and are used as paths by other species as well as hippos. The pastures are not necessarily the same throughout the year. Although hippos travel more in the dry season in search of grass they are able to extend their range much further in the wet season by making use of temporary wallows for daytime resting. These temporary wallows are very important in determining the carrying capacity of the region for hippos as they allow the animals to visit areas which they would be unable to exploit if they were obliged to trek many kilometres to and from their home lake or river every night. These wallows extend the range by about 7 km in Uganda, i.e. up to 10 km from permanent water. Most of the hippos that use the temporary wallows are males but it is unlikely that they include territorial bulls as they would thereby forsake their territories. The paucity of females may be due to their reluctance to walk too far if they have calves that could become easily tired.

It used to be thought that hippos visited their grazing grounds only every other night but this has not been confirmed through detailed observation. Circumstantial evidence comes from the low average weight of the stomach contents relative to the total body weight, which is only about half of what would be expected from comparisons with other large mammals (p. 35). A more likely explanation, however, is that the life style of the hippo conserves energy and does not require so much fuel. Further evidence for the alternate night-feeding hypothesis is the frequent presence of hippos in the water after dark but these may be individuals that have already grazed. It seems that the hippo feeds in bouts for relatively short periods at a time and rests in between, presumably to allow time for digestion. If it is some distance from water, the satiated hippo moves into a bush to rest but if it is close to its home river or lake, it can easily slip back into the water for a temporary rest. Whatever form their nocturnal activity may take, most hippos are back in their wallows well before dawn. Hence it is not surprising that they have been reported in the water at night.

Table 4.1   The distances (km) covered by hippos in travelling to their grazing grounds, Lundi River, Gonarezhou National Park, Zimbabwe. Data for wet and dry seasons in 1981. (After O'Connor & Campbell, 1986.)

| | Total Distance | | Farthest Distance from Water | |
	Dry Season	Wet Season	Dry Season	Wet Season
North Bank				
Mean	★	0.85	0.71	0.34
Maximum	★	2.35	2.12	1.15
Median	★	0.58	0.53	0.19
South Bank				
Mean	1.64	1.70	0.71	0.52
Maximum	1.90	2.70	0.84	1.16
Median	1.85	1.69	0.81	0.47

★Only partial records available.

Grazing densities

It is difficult to compare the densities of hippos in various regions because the total area to be considered is not always easily identifiable. One measure is the number of hippos per linear distance along a river bank or lake shore but this does not tell us much about the overall density as those stretches that do not contain hippos are often excluded from the calculations. The number of hippos divided by the area of a national park or reserve is not a meaningful measure either, as much of the region may not be suitable for hippos. Thus the 1978 km^2 Queen Elizabeth National Park had a post-cull population of about 12 000 hippos but a simple calculation of the density as six hippos per square kilometre ignores the fact that much of the park consists of forest and thick bush, which are uninhabitable for hippos. If only those regions suitable for hippos are included, the density in one area rises to as many as 28 hippos km^{-2}. In Uganda's other large national park, Murchison Falls, hippo grazing densities along the River Nile ranged from 9.4 to 26.5 hippos km^{-2} (Laws *et al.*, 1975). There were also temporal differences as densities calculated from two counts made in 1964 and 1967 respectively, showed marked differences in the same sections of the river. The overall density fell from 19.2 km^{-2} to 16.0 km^{-2} over the three-year period.

These densities are among the highest in Africa and in terms of biomass density – weight per unit area – Uganda holds the world record, or at least it did before the political troubles of the 1970s wiped out large numbers of the heaviest mammals. Using a modest 1000 kg for the average weight of a hippo (taking into account the calves as well as the large bulls), the density in one of the study areas in the Queen Elizabeth National Park amounted to 27 990 kg km^{-2} i.e. nearly 28 tonnes (Field & Laws, 1970). If all large mammals are included, the biomass density rises to 36 511 kg km^{-2}. To put this in perspective, the equivalent density in the Serengeti National Park in Tanzania, a region famous for its abundant wildlife, was only 6300 kg km^{-2} (Stewart & Talbot, 1962).

The biomass densities in the Murchison Falls National Park were equally impressive. They were estimated to be 21.1 hippos (21 100 kg) km^{-2} south of the Nile and 26.9 hippos (26 900 kg) north of the river (Laws *et al.* 1975). These hippo densities were some six or seven times those of the elephants in the same regions. These numbers were considered to be excessive by Laws and his colleagues, who maintained that the grazing activity of the hippos was the cause of habitat degradation. Possible management solutions to this problem will be considered later.

There are few estimates of hippo densities outside Uganda but Viljoen & Biggs (1998) give some figures for the Kruger National Park in South Africa. Densities over an 11-year period on three of the most populous rivers were 1.32 km^{-2} on the Olifants River, 1.16 km^{-2} on the Sabie River and 1.01 km^{-2} on the Letaba River. These figures were based on the number per length of river and converted into densities on the assumption that the grazing area extended up to 3.2 km from either bank. Other measures of hippo numbers per kilometre of river, but not converted to grazing densities, are available for the Luangwa River in Zambia. A figure of 18.8 km^{-1} is given by Ansell (1965), one of 73.8 km^{-1} by Norton (1988) and one of 38.0 km^{-1} by Tembo (1987). The wide deviation between the 1987/88 figures does not give one much confidence that any of these values were typical. Other estimates include 16.1 km^{-1} on the Mara River in Kenya (Olivier & Laurie, 1974a) and between 114 and 145 km^{-1} on rivers in the Benoue National Park in Cameroon (Ngog Nye, 1988). Too much significance should not be attached to these data, for measurements made over short stretches of river are meaningless owing to the habits of hippos to congregate in restricted areas. Only those counts which include more or less the whole length of a river are at all reliable.

Dispersal

All species must have a dispersal phase in their life history to avoid the deleterious effects of inbreeding (Greenwood, 1983). This usually takes the form of one or other sex leaving their natal area on reaching maturity and moving over a long distance before settling down to find a mate. In about 72% of the 67 mammal species reviewed by Greenwood (1983), it is the male that emigrates. Female dispersal is rare, being found in only some 8% of species, including, interestingly enough, our own, for in most cultures it is the wife who goes to live in her husband's village. Female dispersion also occurs in our closest relatives, the chimpanzee and gorilla. Both sexes disperse in the remaining 20% of species, which include the plains zebra and the lar gibbon. Nothing is known about the hippo in this respect although on the balance of probabilities, it should be the male that disperses. The only evidence, circumstantial though it is, suggests that this may not be the case. This is based on the story of the extraordinary wanderings of a female hippo in South Africa during the late 1920s. The account is given in a book by Hedley Chilvers, which was probably published in 1931 although no publication date is given.

The hippo, named Huberta, was born in St Lucia Bay in Zululand and was first noticed in 1928 when she crossed the Tugela river into Natal. She then followed a route south over the next two and a half years which took her almost to Grahamstown some 550 miles away. She was easily recognisable because she passed through areas where hippos do not normally occur. She reached Durban in June 1929 and spent some time there. Another town visited on her journey was Port St John's, where she was seen in the market square at 3 o'clock in the morning. By now she had been declared royal game and given legal protection but she was, in any case, left alone by the local people, who thought she was a reincarnation of a witch doctor. She met her end in April 1931 on the Nahoon River, near Berlin, where she was shot by the sons of an illiterate farmer, who were later charged with destroying royal game without a licence and fined £25 each.

No satisfactory explanation has been given for such a long journey for it certainly did not lead to Huberta finding a mate. It may be that other hippos make similar movements but they would not be noticed if they moved to regions containing existing populations of hippos. Satellite tracking of radio-collared animals seems called for if further progress is to be made.

THE BEHAVIOUR OF *HEXAPROTODON LIBERIENSIS*

Very little is known about the behaviour of the pygmy hippo in the wild. The problems associated with its study were alluded to at the beginning of this chapter, where the triple difficulties posed by a rare, nocturnal, forest creature were explained. Although there have been very few studies of the species, it is known that its behaviour differs from that of the large hippo in a number of significant ways. In the first place, the pygmy hippo is much more overtly a solitary animal and is not found in groups like the larger species. Most observations are of single animals although pairs and threesomes are not unusual. It is not clear whether two animals together constitute a monogamous pair but it is more likely that the male is promiscuous. According to Lang *et al.* (1988), the home range of a male is about 185 ha and covers those of several females but it is not known whether it is a territory in the strict sense of being a defended area. It probably is not, for there does not appear to be any attempt to exclude other hippos. When two animals meet, they tend to ignore each other rather than threaten or fight. The home range of a female varies from 40 to 60 ha and to overlap the ranges of other females. If a third animal is present in a group, it is probably a youngster.

The pygmy hippo is said by Robinson (1970) to be more like a tapir than the large hippo in its ecology and social behaviour. It is mainly nocturnal although not exclusively so and it is often abroad in the late afternoon. It spends most of the day hidden in swamps or amongst the aquatic vegetation of rivers. It has long been thought to occupy holes in river banks and indeed Schomburg (1912), the first person to capture a pygmy hippo alive, maintained that the species occupied hollows and tunnels in river banks.

Some scientists felt that this was over-imaginative and no further reports were forthcoming until Robinson (1981) gave a detailed description of one such tunnel in Grand Bassa County, Liberia (Fig. 4.1). Although it had been abandoned in the previous year after the roof had collapsed in two places, local hunters, who had found the den, insisted that they had seen a pygmy hippo emerge from it on a number of occasions. Robinson inspected the structure in February, 1968, during the dry season, when the water level was at its lowest, and saw two entrances, which were between the erosion-exposed roots of a large *Parkia* tree. About two-thirds of the entrance holes were above the stream level at the time but they would have been completely submerged in the wet season. The entrance tunnels led into a chamber measuring 4.5 m in length, 3.1 m in width and 0.6 m in height. It is not clear whether the hippo

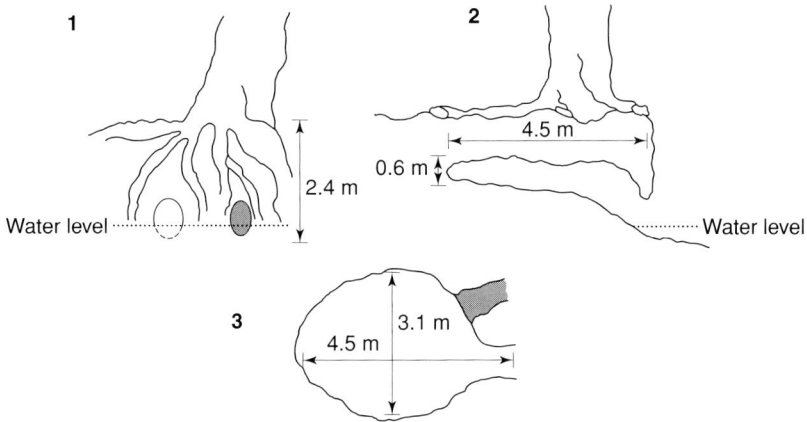

Figure 4.1   The denning structure used by a pygmy hippopotamus in Liberia. 1 – External view showing open and collapsed (shaded) entrances. 2 – Lateral cutaway view. 3 – Plan of den from above showing open and collapsed (shaded) entrances. From Robinson (1981).

had excavated the chamber itself or had merely enlarged an existing hole that had formed naturally between the roots of the tree. In any case the observation provides good evidence of the denning behaviour of pygmy hippos. It might seem surprising that a hippo should burrow underground but not when the tunnelling behaviour of other members of the Suiformes, such as warthogs, is considered. The hippo changes its sleeping quarters about once or twice a week.

Even if it does not always occupy a burrow, the pygmy hippo spends the day resting in or near water and emerges at dusk to feed, just like its larger relative. It too follows well-defined game trails, which it marks with faeces dispersed by vigorous tail-wagging, again like the behaviour of the large hippo. These trails become like tunnels in the forest vegetation and like canals in the swamps. In one area, the density of the hippo trails reached 32 paths km^{-2}.

The pygmy hippo spends about six hours a day in feeding, usually between mid-afternoon and midnight. It differs from its relative in what it eats. Although it does take grass, of which there is little in forests, the principal food appears to be the leaves and roots of forest plants as well as fallen fruits and ferns (see page 84). This is a higher quality diet than the grass eaten by *H. amphibius* and needs less digestion.

CHAPTER 5

REPRODUCTION IN HIPPOS

Female common hippo with new-born calf

REPRODUCTION IN *HIPPOPOTAMUS AMPHIBIUS*

Breeding seasons

Much of our knowledge of seasonal reproduction in *H. amphibius* derives from the examination of specimens culled in the early 1960s in Queen Elizabeth National Park, western Uganda, following the over-grazing by hippos described on page 89 (Laws & Clough, 1966). Between November 1961 and October 1962, reproductive data were obtained from 592 of the hippos that were shot. The material collected from females included foetuses in their membranes, ovaries, segments of uterus and vagina, placentas and mammary glands. Male organs included testes, epididymides, prostate glands, Cowper's glands and seminal vesicles. Material of reproductive relevance collected from both sexes comprised adrenals, thyroids and pituitaries. The samples were preserved for later examination in the laboratory. In addition several complete reproductive tracts were dissected out and examined either in the field or in the laboratory.

Another important reproductive field study involving 337 culled animals was carried out on the Luangwa River in eastern Zambia, where about 1000 hippos were shot between 1965 and 1970, again in response to perceived over-abundance (Marshall & Sayer, 1976; Sayer & Rakha, 1974). Further observations were made during extensive culling operations that were carried out in the Kruger National Park, South Africa during the 1970s. The only previous culling in the park was an experimental operation during the 1964 drought when 104 hippos were taken out (Pienaar *et al.*, 1966). The methodology for examining the animals in Zambia and South Africa was similar to that employed in the Ugandan project.

Male hippos in Uganda have active spermatozoa in their reproductive tracts throughout the year and are capable of fertilising the females at any time. There is, however, a marked seasonality in the female. Births can take place at any time of the year in Uganda but there are peaks associated with rainfall. Seasonality was not apparent in the proportion of pregnant females in the sample but by applying a growth equation (Hugget & Widdas, 1951) to the length of foetuses, the periods of peak conceptions and births could be obtained. Recalculation of data given in Laws & Clough (1996) showed that peak conceptions occurred towards the end of the wet season with a peak in births towards the beginning (R.M. Laws, *in litt.*).

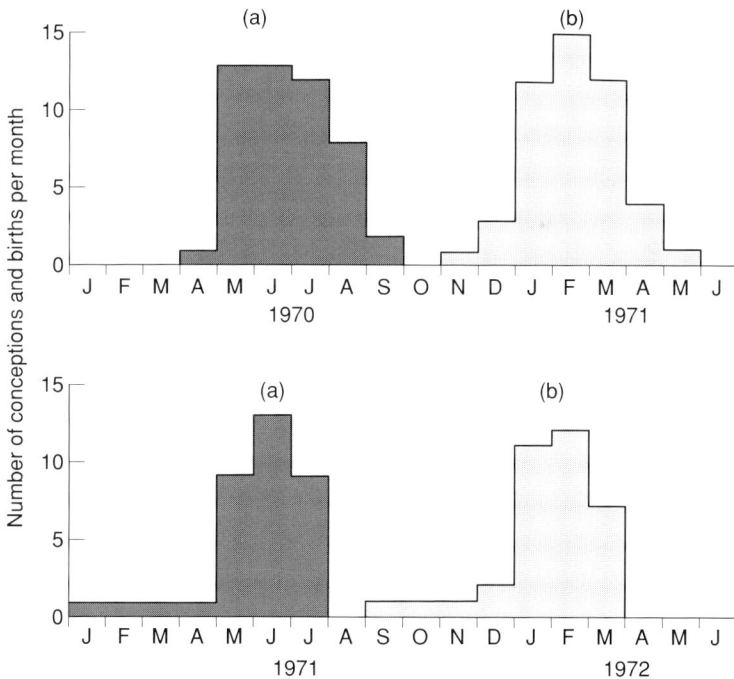

Figure 5.1 Reproduction in the common hippopotamus in Zambia some 10–15 degrees south of the equator under a unimodal rainfall regime. Births are seasonal, with none occurring in some months. a – conceptions, b – births. From Marshall & Sayer (1976).

Using similar techniques, Marshall & Sayer (1976) found a much more sharply defined reproductive seasonality in the Zambian hippos (Fig. 5.1). Conceptions occurred throughout the year, although at very low levels in the dry season, but births were restricted to the wet season and rose to a peak at the height of the rains.

Smuts & Whyte (1981) could not assess seasonality in the male hippos in South Africa because specimens were collected over only a limited period but they found seasonality in the female with a marked peak in births around the time of maximum rainfall (Fig. 5.2).

It is normal for herbivorous mammals to correlate their births with the onset of the rainy season as the quality of the foliage is greatest at that time of year. This is particularly true of grasses, which produce proteins and soluble carbohydrates within three or four weeks of the first showers.

The greatest drain on the female's resources occurs, not during gestation, but during lactation (this is obvious when one considers the birth and weaning weights of the young). The relative growth rates of the foetus and of the new-born calf reflect this physiological constraint with a two-thirds increase in the weight to length ratio occurring after birth. The wet weights of the stomach contents in the Ugandan

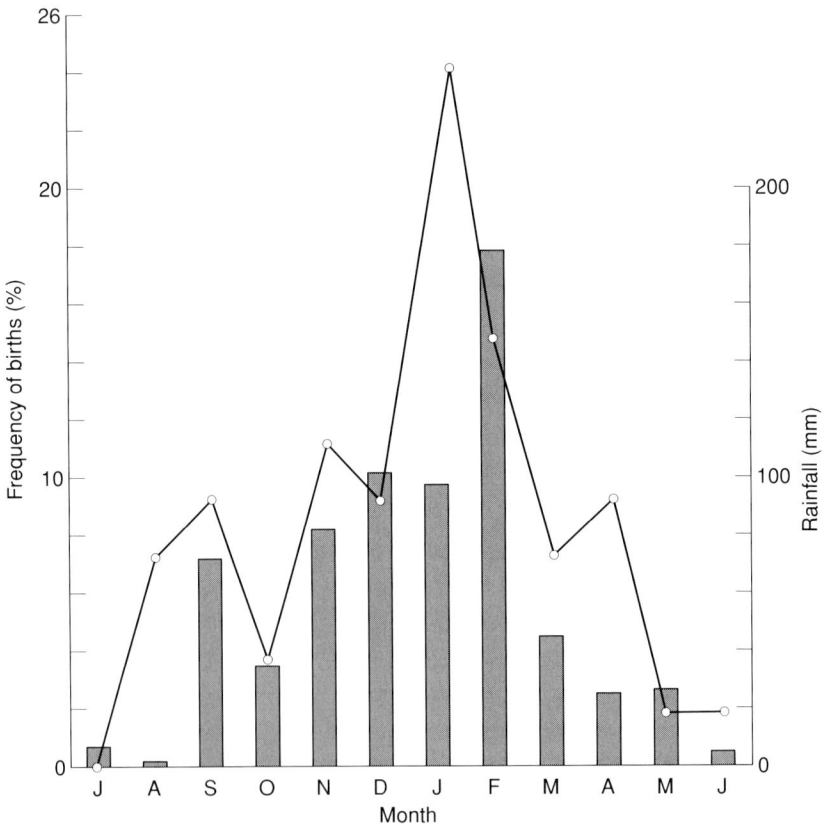

Figure 5.2 Reproduction of the common hippopotamus in South Africa showing a marked correlation between births (graph) and rainfall (histogram). From Smuts & Whyte (1981).

sample show that lactating females, with a stomach fill of 214 kg, feed more intensively than pregnant animals, whose stomach contents weigh only 183 kg. Hence, for a grazer like the hippo, calves that are born a few weeks into the wet season are likely to survive in better condition than those born at other times of the year. Nevertheless, nearly a third of the births recorded by Laws & Clough occurred when the monthly rainfall was less than 5 cm.

It must be emphasised that much of the information derived from the Uganda hippos refers to a population living close to the Equator, where there are two dry seasons a year, each only three months long, so the period of deprivation is quite short. Some rain may even fall in the dry season. Where there is a prolonged dry season, as in Zambia and South Africa, breeding in the hippo is a much more seasonal phenomenon. It should also be pointed out that the study hippos in Uganda were living under conditions of perceived over-abundance. Indeed, that was the reason why they had been culled in the first place. Consequently, the hippos may have been under stress, which could have affected their reproduction, and the conclusions do not, therefore, necessarily apply to all hippos throughout Africa.

The sexual cycle

As mentioned above, sexual cycles in the male hippo do not occur, at least in Uganda, and a mature male is fertile throughout the year. In this, it is typical of most large mammals. This is not the case with the female, which, in common with the majority of eutherian mammals, undergoes fluctuating periods of sexual activity. These oestrous cycles, as they are called, are related to changes within the ovary, which can be traced by dissection. A mammalian egg is a tiny structure that develops from a cell of the germinal epithelium, which surrounds the ovary. The hippo is monotocous, except in the case of twins, and when the cell divides, one of the daughter cells, known as the primary oogonium, passes into the ovary and becomes surrounded by follicle cells. It then grows, without further division, into the oocyte under the influence of the follicle stimulating hormone. The follicle cells pass fluid into the cavity, or antrum, of the developing follicle, which consequently increases in size to become a Graafian follicle. Eventually the distended follicle bursts, releasing the egg into the reproductive tract, where it becomes embedded in the uterine wall after fertilisation. The process is known as ovulation and in most mammals is accompanied by behavioural changes known as oestrus. The mature Graafian follicle develops into a corpus luteum, the "yellow body", which becomes an endocrine organ helping to control the events of pregnancy. It degenerates soon after birth of the calf into a whitened body, the corpus albicans. The presence and stage of development of these bodies in the ovary provide a useful indication of the point in the oestrous cycle that the female has reached.

In addition to the corpus luteum of pregnancy, there are, in the hippo, a number of accessory corpora lutea. These are solid structures which could have arisen from ovulations during pregnancy or from the complete luteinisation of follicles. The number of accessory corpora lutea increases from an average of 0.3 in early pregnancy to 7.5 at birth (Fig. 5.3). The production of accessory corpora lutea is shared with other species, such as the horse and elephant, but the peculiarity in the hippo is that the luteinisation process is progressive throughout pregnancy.

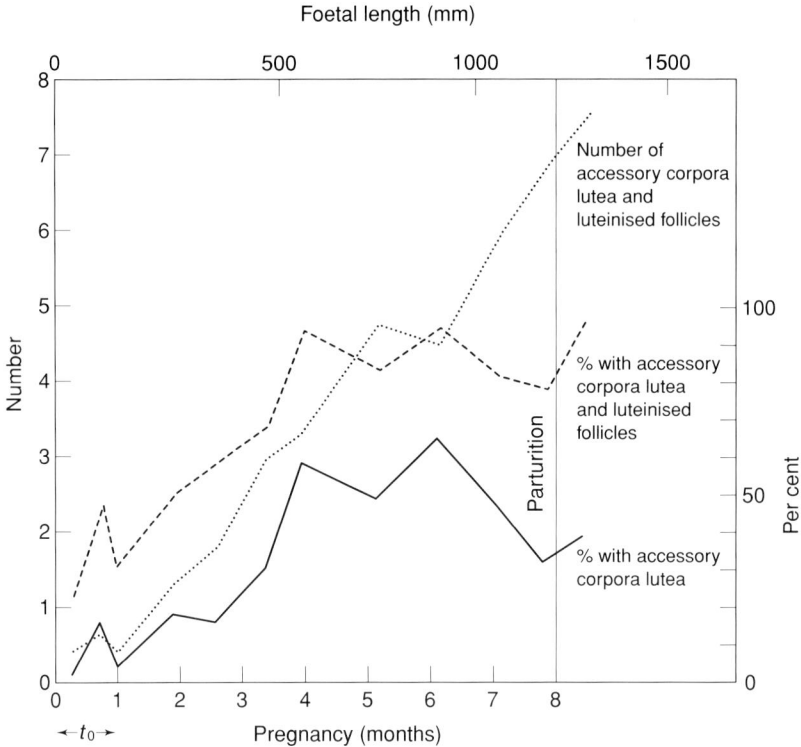

Figure 5.3   Changes in the ovaries of the common hippopotamus during pregnancy. Unlike most mammals, the number of accessory corpora lutea increases almost linearly and luteinisation continues throughout pregnancy. From Laws & Clough (1966).

Calving interval

The hippo is polyoestrous, i.e. it will repeat the cycle more than once in the year, so if an egg is not fertilised after ovulation the female will come into oestrus again. If it becomes pregnant, the oestrous cycle ceases and the female moves to a period of lactational anoestrus. Laws & Clough (1966) postulate a two-year sexual cycle based on a six-month oestrous cycle, an eight-month gestation and a ten to twelve-month lactation period. This means that a female could produce a calf about every two years. Some mammals, however, come into heat soon after parturition in what is known as a *post partum* oestrus and there is evidence that this can happen in the hippo, for about 25% of the females examined by Laws & Clough were both pregnant and lactating. Hence the calving interval could be less than two years although it is possible that some of the pregnant, lactating females had recently lost their calves, in which case they would return to oestrus. In the absence of a successful mating during lactation there is usually a four-month period of anoestrus.

Smuts & Whyte (1981) calculated the mean calving interval of their hippos in the Kruger National Park to be 21.8 months based on the ratio of pregnant to non-pregnant females, and assuming a gestation period of eight months. This suggests a two-year sexual cycle, as assumed by Laws & Clough. Smuts and Whyte, however, question the latter's conclusion and recalculate their data, using a different method, to produce a calving interval of 32.5 months for the Ugandan hippos, suggesting a three-year sexual cycle. It is not clear why the cycle should be a year longer in Uganda than in South Africa.

Observations on two captive hippos by Verheyen (1954) showed that the females came on heat 37 and 42 days respectively after giving birth, suggesting that these periods represented the durations of the oestrous cycle. In six other cases the period between parturition and oestrus was calculated from the average interval between successive births, which was 292 days. Assuming a mean gestation period of 240 days, the interval would, therefore, have been 52 days. If the first two cases are included, the mean period becomes 49 days. These observations are subject to several errors (e.g. the female may not immediately return to oestrus on giving birth) and they cannot be used to define the length of one oestrous cycle but they do not rule out the assumption that it is of the order of one month, as in many other large mammals.

Calving rates

The calving rate is measured as the percentage of females that produce a calf each year. Smuts & Whyte (1981) found a conception rate of 36.7% for the hippos in the Kruger National Park in South Africa. This is higher than other published figures, which vary from 20 to 36.75%, but it is low compared with the rates in other large herbivores with a similar gestation period. Pregnancy rates were 27% in Uganda (Laws & Clough, 1966). As not all foetuses will survive to term, the actual calving rates will be rather lower than these figures.

Mating

Mating in the hippo takes place in the water, with the female submerged for most of the time. It is a prolonged and noisy affair during which only the male can be seen, rearing up out of the water, and the only indication of what is taking place is the periodic emergence of the female's head, presumably to draw breath.

Gestation

The gestation period of the hippo is known fairly accurately from observations of zoo animals and is about 240 days, i.e. not quite eight months. This is short for such a large animal; for comparison, the somewhat smaller black rhinoceros has a gestation period of about 15 months.

Birth

Where the female hippo gives birth is debatable. Some births certainly occur in the water but whether very many take place there is disputed. There is no problem for a young hippo over a water birth any more than there is for the birth of a whale. The water is usually shallow and the baby simply does not breathe until it reaches the surface. The female separates herself from other hippos when about to give birth and keeps away for a couple of weeks, during which time she is fiercely defensive of the calf and can be dangerous to people. She is also aggressive towards other hippos, whether they be territorial males or her own grown offspring. This has the effect of preventing the young calf from imprinting on any hippo other than its mother.

The large hippo breeds readily in captivity and several captive births have been described, of which the most detailed is probably that by Senior & Tong (1963). This took place in shallow water at Whipsnade Park on 2nd July 1961. The calf was born in a breech presentation (hind feet first) after a short labour of less than an hour although the female had shown signs of the impending birth over the previous month.

The birth weight of *H. amphibius* was calculated by Laws & Clough (1966) from a study of foetal growth to be 50 kg although one newborn calf was as light as 25.45 kg. Presumably it was premature or possibly one of twins. Twins do occur although not very frequently – only two sets were recorded in the sample of 276 specimens examined by Laws & Clough. While I was working in Uganda, a young hippo calf, which was brought to me by a fisherman, was only half the birth weight of a normal hippo and I suspect that it was one of a pair of twins. The presumption is that the mother was satisfied with just one of the calves and neglected to notice that the other had become separated from her. The heaviest foetus recorded by Marshall & Sayer (1976) from a sample of 84 Zambian females was 44 kg, which compares well with the Ugandan estimate. Smuts & Whyte (1981) concluded from a review of the literature that the average birth weight was 42.14 kg.

The mean body length at birth in the sample studied by Laws & Clough was calculated from the growth rates of the foetuses to be 127 cm.

Lactation

The female hippo has two mammary glands situated in the inguinal region. The glands are nearly 50% heavier in a lactating female than in one that is neither pregnant nor lactating. Suckling takes place either on land or in the water with the mother usually lying on her side. The calf may also suck under water and I suspect that milk is injected into its mouth by muscular action on the part of the mother, as is found in some cetaceans. I base this belief on the behaviour of the abandoned calf mentioned above, which coiled its tongue into a groove when presented with its milk bottle. This made it very easy to "inject" milk into its throat. Under natural conditions, the calf's tongue is presumably wrapped around the teat.

The percentage of lactating females in Uganda (Laws & Clough, 1966) was 59.9% with a further 6.2% being both pregnant and lactating. The corresponding values for South Africa (Smuts & Whyte, 1981) were 78.0% and 20.2%. The duration of lactation is not known but it is probably of the order of one year. As with most mammals, there is no abrupt cessation of suckling and the calf continues to suck long after it has learnt to graze. Laws & Clough found that calves as young as six to eight weeks had considerable quantities of grass in the stomach. In general, however, weaning can be considered to occur between six and eight months, with most calves being fully weaned by 12 months of age. This is not to say that a calf older than this would not attempt to suck from its mother but it would most likely be repulsed.

Smuts and Whyte (1981) suggest that hippos cope with poor environmental conditions by extending the lactation period and, possibly, by calves being suckled by more than one cow. The latter is highly unlikely unless the two animals are close relatives, as in the case of elephants cited in support by Smuts and Whyte. The evidence, such as it is, is circumstantial, and is largely based on there being more lactating females than calves in samples of shot animals.

Sex ratios

The sex ratio at birth is probably 1:1, as would be expected. The figures for the Ugandan foetuses examined by Laws & Clough (1966) were 116 males to 121 females a ratio of 1:1.04, which does not differ significantly from unity. Bere (1959), from an earlier analysis of the same population, found an exactly similar ratio with 92 foetuses. Smuts & Whyte (1981) also found a sex ratio of precisely 1:1 in their sample of 32 foetuses from the Kruger National Park and Suzuki (1997) one of 1:0.71 in a smaller sample of 24 from the Luangwa Valley, Zambia.

Adult sex ratios are more difficult to assess because it is difficult to obtain a representative sample. Most information comes from culled animals but it is known that culling does not select a random sample of the population. Animals shot in wallows are more likely to be males and culls taken from "schools" along shorelines have a preponderance of females.

Not much credence, therefore, can be attached to adult sex ratios although there are several figures quoted in the literature. The overall figure for the 463 hippos of all ages culled in the Kruger National Park in the four years between 1974 and 1976 was 158 males and 305 females or 1:1.93 (Smuts & Whyte, 1981). Marshall & Sayer (1976) found a sex ratio (males:females) in the Luangwa Valley of 1:0.98 in 1970 (375 in sample) compared with 1:1.56 in 1971 (210 in sample). The smaller proportion of males recorded in the second year was attributed to the greater effort made in 1970 to collect scattered hippos, which are more likely to be males than those in herds. An even balance of the sexes (1:0.92) was found by Sayer & Rakha (1974) in their sample of 336 hippos, also culled in the Luangwa Valley. In a much bigger sample, Laws & Clough (1966) found an almost exact sex ratio of 1270 males to 1271 females. Their data confirmed the inequitable distribution of the sexes according to habitat. Thus in small wallows there were 313 males and 110 females whereas on the lake shore the proportion was reversed with 600 males and 821 females. The proportions for hippos shot on land at night, which might be expected to be more random, was 170 males to 210 females (1:1.24).

The latter is probably the most reliable sex ratio and agrees with theoretical expectations. In most mammalian species with seasonal breeding, the life expectancy of males is less than that of females mainly because the males are preoccupied with sex at a time of year when they should be following the example of females and feeding to put on fat to last through the lean period, whether winter or dry season. Deaths in the males are more likely to be due to inadequate nutrition than to fighting. Although the evidence is not conclusive, it is probable that male hippos die earlier than females. Certainly the percentage of females in the sample collected by Smuts & Whyte (1981) shows a steady, if erratic, increase with age.

Age at puberty

The age at puberty in the male is based on a number of criteria including the presence of viable spermatozoa in the epididymis, testes weight, diameter of testicular tubules, percentage of open tubules and the percentage of interstitial tissue in the testis (Fig. 5.4). The onset of puberty was estimated by Laws & Clough for Ugandan hippos from an examination of the gonads. In the male the combined weights of the testes and epididymides from 442 hippos were plotted against age. These organs were considered to weigh a minimum of 400 g in a fully sexually mature male and this weight was reached by some animals at between six and seven years of age. Others did not attain this level until they were 15 years old. Allowing for the statistical ranges attached to these figures, puberty in the male was assumed to lie between six and 16 years of age although the upper limit was considered to be anomalous and a better estimate would be six to 13 years. A rather lower limit results from records of the presence or absence of spermatozoa in the epididymis. On this criterion, the age at puberty lies between four and 11 years. Taking all factors into consideration, Laws & Clough concluded that the average age of puberty in the male is about 7.5 years.

Sayer & Rakha (1974) examined 176 male hippos that had been culled in Zambia, where the rainfall pattern is very different from that in Uganda in that there is only one rainy season, which lasts from October to May with a peak in June (Fig. 5.5). They weighed the testes, less the epididymides, and showed that there was a sharp increase in weight between age classes V and VI, suggesting that puberty was reached at about six years of age (Fig 5.6). A direct comparison could not be made with the Ugandan measurements, since the weights of the epididymides were not included,

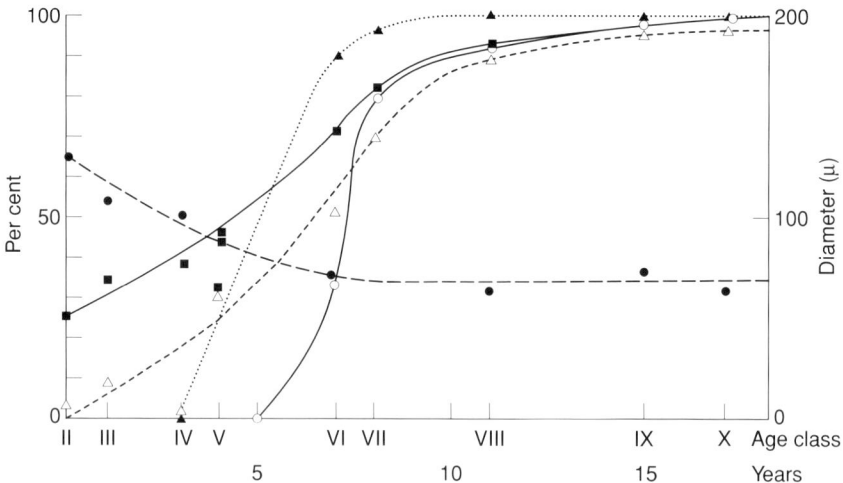

Figure 5.4 Puberty in the male common hippopotamus based on various measurements of the reproductive organs in hippos shot in Uganda. From Laws & Clough (1966), who conclude that puberty occurs between 6 and 13 years of age. Key: ○—○ % mature based on testis and epididymis weights; ■—■ mean testis tube diameter; △---△ % with open testis tubules; ▲····▲ % with sperm in epididymis; ●---●% of interstitial tissue.

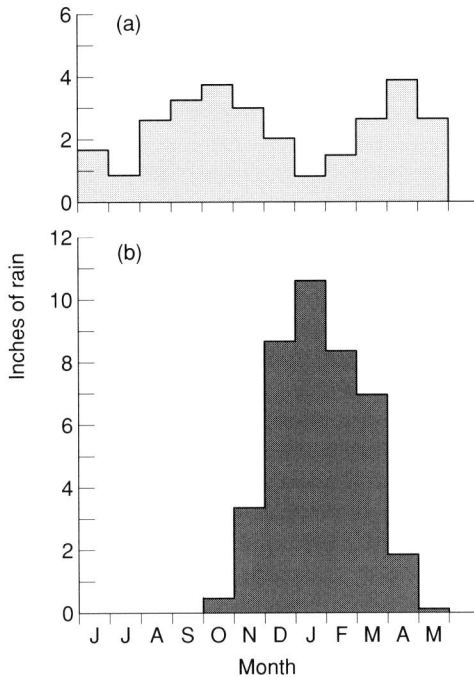

Figure 5.5 Seasonal distribution of rainfall in Uganda (a) and Zambia (b) showing that a bimodal distribution (two peaks) on the equator becomes a unimodal distribution (one peak) 14 or so degrees south. From Marshall & Sayer (1976).

but the combined weights of the testes approached 400 g (Laws & Clough's criterion) at approximately eight years of age, close to the figure of 7.5 years deduced for the Uganda hippos.

An interesting side line revealed by Sayer & Rakha's data is that the testis continues to grow throughout life. This may be related to the fact that body weight in the male hippo also continues to increase with age (p. 10). It is likely therefore that the male hippo is fertile throughout life. Only one male in Laws & Clough's sample had a testis weight that was unusually low but, even so, the testes appeared to be functional.

Smuts & Whyte (1981) found that spermatogenesis in the South African hippos began at as early as two years of age but it was not until they were six years old that all males had active sperm in the epididymis. The combined testes weight then was 266 g. The 400 g threshold used by Laws & Clough as the criterion for puberty was reached at around seven years. There is, therefore, unanimity in the various studies for placing the average age of sexual maturity in the male hippo at between seven and eight years.

The age of puberty in the Uganda females was deduced from an examination of ovarian activity (Laws & Clough, 1966). Females were assumed to be mature if one or more corpora lutea or corpora albicantia were present. The size of the Graafian

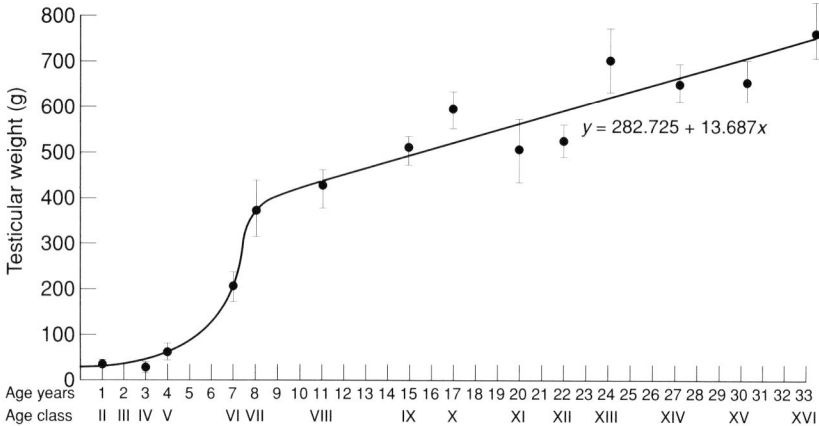

Figure 5.6 Puberty in the male common hippopotamus based on testes weights of hippos shot in Zambia. The sharp increase in the average weight between age classes V and VII suggests that puberty is reached at about 6 years of age. From Sayer & Rakha (1974).

follicles was also taken into consideration. These follicles grow bigger as puberty approaches and appear to reach their full size some three years before sexual maturity. On the basis of these measurements and taking other modifying factors into account Laws & Clough put the age of female sexual maturity at about nine years with a range of seven to 15 years (Fig. 5.7). Fifty per cent of the females were mature at 9.5 years.

Sayer & Rakha (1970) used criteria similar to those of Laws & Clough in their examination of 161 culled female hippos from Zambia. They related the onset of puberty to follicular size and considered a female to be mature if it was pregnant or lactating. They concluded that puberty was reached between seven years of age, plus or minus a year, and 17 years, plus or minus three years. By about 13 years, 50% of the females were either pregnant or lactating. These ages are not radically different from those estimated by Laws & Clough for Uganda hippos, although they are a little older (Fig. 5.8).

Further culling of hippos in the Luangwa River took place in 1995 and 1996, when the population had about doubled in size despite a massive die-off due to anthrax in 1987. In 1995, 507 hippos were shot, of which nearly 400 were examined scientifically (Suzuki & Imae, 1996). The percentage of lactating females in relation to age was determined and it was found that lactation started at about 11 years of age, with 50% of the sample lactating by the age of 17. This is similar to the results Sayer & Rakha (1976) obtained 25 years earlier. One might have expected puberty to be delayed in view of the increased density but the population might still have been below carrying capacity. The lack of a difference could, however, be an artefact of the age estimation technique, for Suzuki & Imae report that there could be a two-year ambiguity in their assessments. A further 234 hippos were shot in the following year in regions that had not been used for the 1995 cull (Suzuki, 1997) but there do not seem to have been any significant changes in the reproductive variables.

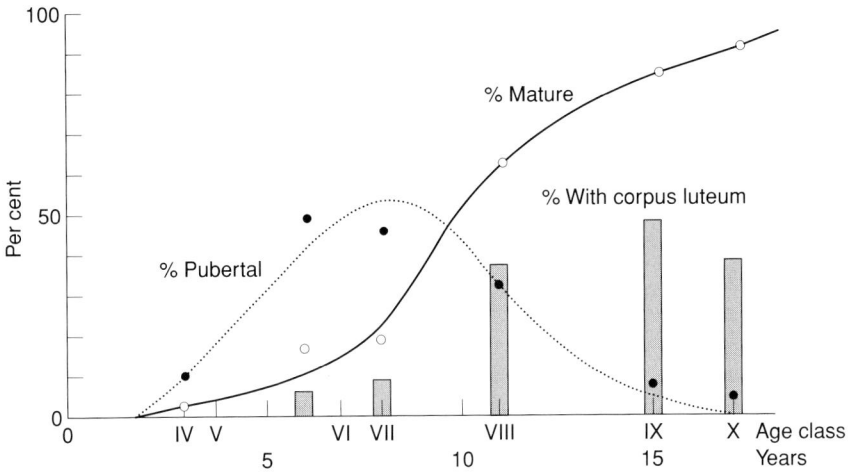

Figure 5.7 Puberty in the female common hippopotamus based on ovarian activity in hippos shot in Uganda. From Laws & Clough (1966), who conclude that puberty occurs at a mean age of 9 years with a range of 7 to 15 years.

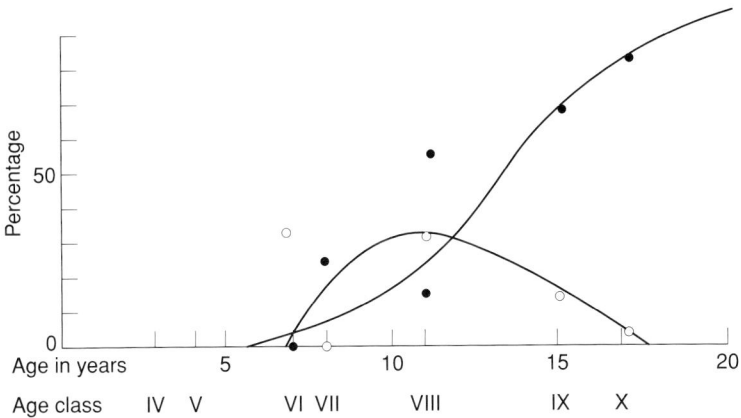

Figure 5.8 Puberty in the female common hippopotamus based on hippos shot in Zambia. Open circles – pubertal; closed circles- mature. A female was considered mature if it was pregnant or lactating and pubertal if it had one or more Graafian follicles with a diameter of at least 12 mm. From Sayer & Rakha (1974).

The average age of sexual maturity, based on a 50% pregnancy rate, was nine to ten years. This is close to the 50% figures for Uganda (9.5 years) and Zambia (11 years). The age at which *all* females are pregnant, however, does vary between the three regions. In the Kruger National Park it was 11 years but the 100% level was not reached until 20 years of age in Uganda and Zambia. This suggests that conditions in the latter two countries were more severe than in South Africa. This is not surprising as far as Uganda is concerned as it is known that the hippos there were living under crowded conditions and it seems likely that this was also true of the Zambian population. It might be expected from density-dependent considerations that overcrowding would retard reproductive development.

Some of the females examined were sexually mature at much younger ages than the averages mentioned above. Laws & Clough (1966) found that one three-year-old female was sexually mature. The youngest pregnant female in the Kruger samples (Smuts & Whyte, 1981) was only six years old and so must have been sexually mature at five years of age.

The results of these various estimates of the onset of puberty are summarised in Table 5.1.

These ages have been challenged as being far too high by Dittrich (1976) on the grounds that zoo births have occurred at much lower ages than those estimated from field studies. That young male hippos can be sexually precocious was demonstrated by the three- to four-year-old male in Antwerp Zoo, which successfully fathered a calf (Bourlière & Verschuren, 1960). Goss (1960) reported a birth at Cleveland Zoo to parents both of which were only three years old at the time of conception.

Females in captivity have been known to breed at a very early age. The youngest recorded by Dittrich (1976) was two weeks short of three years on giving birth, which, based on a gestation period of eight months, means that she was only 2 years and 3.5 months old when she was successfully mated. The average age at which females first conceived in 24 zoo births around the world recorded by Dittrich can be calculated as about 4 years 7 months with a range of 2 years 3.5 months to 7 years 10 months. Similar calculations for eight captive male hippos show that the average age at which they first successfully fathered a calf, i.e. were fertile, was 4 years 7 months, with a range of 2 years 6 months to 7 years 3 months. As Dittrich points out, sexual maturity could well have occurred at a much lower age in the case of the older animals as a zoo may not have had a male available when the female was first in oestrus.

Table 5.1 Estimates in years of the age of puberty in the common hippo from dissections of wild animals shot in culling operations.

Estimate (years)	Region	Criteria	Authority
MALES			
6–16	Uganda	Testes weights	Laws & Clough, 1966
4–11	"	Spermatogenesis	"
7.5	"	All factors	"
6–8	Zambia	Testes weights	Sayer & Rakha, 1974
6	Kruger	Testes weights & spermatogenesis	Smuts & Whyte, 1981
FEMALES			
9 (7–15)	Uganda	Presence of corpora & follicular size	Laws & Clough, 1966
7–17	Zambia	Follicular size	Sayer & Rakha, 1974
9–10	Kruger	Lactation	Smuts & Whyte, 1981
11	Zambia	Lactation	Suzuki & Imae, 1996

This discrepancy between the ages of sexual maturity determined for wild and captive hippos is not easy to explain. It may be more apparent than real for the ranges for captive and wild animals overlap. Zoo specimens may be on a better feeding regime. Although it is doubtful that the wild hippos were in particularly poor condition, it is likely that tooth wear in zoo animals is less as they are mostly fed on softer food than is available to wild animals so that they appear to be younger than wild animals of the same age. Dittrich does not accept these arguments and comes to the conclusion that the conversion of Laws' age classes into years is faulty so that animals are younger than the ages assigned to them. Laws (1968a) was able to examine only six lower jaws from known-age hippos, all from zoos and none younger than 15 years, so such errors are not improbable. Dittrich also bases his argument on probabilities by pointing out that the age of puberty in other large mammals is of the order of three to five years both in captive and wild populations. Only the elephant and the great apes mature at an age as high as that suggested for some wild hippos and these species are characterised by a long, dependent "childhood", which is not the case in hippos.

The physiological age of maturity, however, is very different from that of social maturity and a wild male would have to be very much older than the minimum figures quoted above before it could hope to defend a territory successfully.

Unlike males, the females can begin to breed as soon as they are sexually mature and there does not appear to be a significant reduction in fertility with age as is seen, for example, in the female elephant. Of the Uganda females, two in the oldest age class, equivalent on the Laws' criteria to about 43 years of age, had functional ovaries containing young corpora lutea and only one, aged about 40 years, appeared to be in a post-reproductive condition. There was a slight decline in fertility with age in the Kruger females but even so, eight of the nine hippos over 35 years of age were pregnant or lactating.

REPRODUCTION IN *HEXAPROTODON LIBERIENSIS*

Less is known about reproduction in the pygmy hippo than in *H. amphibius* and what little information there is comes mainly from studies of captive animals. This has some advantages in that such variables as gestation period can be measured more accurately but against that, the sample size tends to be low and the artificial conditions of captivity may influence the observations.

Pygmy hippos consort for mating but the duration of the pair bond is not known. In captivity, they breed as monogamous pairs but it is unlikely that they are monogamous in the wild. Lang (1975) and Lang *et al.* (1988) describe matings in Basel Zoo and state that the rut lasts from one to two days but it is not clear whether this refers to the male's period of sexual activity or, more likely, to the duration of oestrus in the female. Copulation takes place either on land or in the water and is similar to that seen in the large hippo although more of the female's body is visible when in the water. Matings occur from one to four times during an oestrous period.

The gestation period of the pygmy hippo is about 188 days according to Lang (1975), although his tabled figures seem to suggest a variation from 196 to 201 days. Some authors maintain that it is longer, e.g. 201–210 days (Anderson & Jones, 1984). At six to seven months, this is not much shorter than the gestation period of its larger relative. The gestation of a male calf takes about one or two days longer than that of a female. As far as I know, a wild birth has never been described but captive births occur either in the water or on land. Care has to be taken to ensure that births do not happen in deep water as drowning has ensued under such circumstances. Lang (1975) provides a series of photographs of a birth, which shows it taking place with the female standing. The calf is born with the hind feet emerging first. As in the common hippo, only one calf is born each time as a rule. The average birth weight of 31 calves in Basel Zoo was 5.73 kg with a range of 4.5–6.2 kg (Lang, *et al.* 1988). Male calves, at 5.90 kg, were slightly heavier than females, which averaged 5.64 kg.

Calves may be produced throughout the year in zoos and there is no indication of seasonality. This would not be surprising in the natural rainforest habitat, because food is plentiful year-round, but seasonal breeding might perhaps be expected under the harsher climatic conditions of a European zoo.

The calf apparently does not follow its mother when she leaves the water to forage but "lies out" in a similar way to that seen in the majority of ungulates. This habit, also known as hiding or, in German, *Ablieger Typ*, is advantageous in that the female, while feeding, does not have the responsibility of looking after the calf, and the calf benefits from the opportunity to rest without the expenditure of energy in walking behind its mother. It is also safer from predation. In such species, the mother returns at intervals and calls the calf out of hiding to suckle it. This tendency to hide the young is seen even in zoos, where, in the pygmy hippo, suckling bouts occur three times a day. As in *H. amphibius*, suckling occurs with the female lying on her side. Even the large hippo shows a tendency to hide young as for the first few days after birth, the mother leaves her calf in the water when she goes to feed although she remains within sight of it.

Weaning occurs at about the same age as in the large hippo, i.e. at six to eight months. Growth continues up to at least the fifth year (Fig. 2.4) and probably well beyond, for the bull "Peppone" at Basel Zoo put on 42 kg between the ages of five and 18.

Sexual maturity is usually reached between three and five years of age (Lang, 1975), which is similar to the age at which the large hippo matures in zoos, lending further support to Dittrich's opinions mentioned above. An exceptionally precocious female, "Farah", in Basel Zoo produced her first calf at the age of 3 years 3 months, having mated when only 2 years 8 months old (Lang, 1975). Two other females were probably more typical, with the first birth occurring at the age of 5 years 6 months in one case and 5 years 8 months in the other. Further births followed at intervals averaging seven, eight and nine months, respectively, for the three females.

HIPPOS AS *K*-STRATEGISTS

In their reproductive habits, both species of hippo show affinity with other large mammals. Animals and plants have been divided into two evolutionary types with those in the first category, *K*-strategists, concentrating on quality rather than quantity in reproduction whereas those in the second category, *r*-selected species, do the reverse. There are certain correlates with each reproductive strategy, e.g. *K*-selected species are large, breed more than once during life, produce few young at each birth event and have a low lifetime reproductive output. Great care is taken over the offspring so that a high proportion is reared to maturity. These are not clear-cut distinctions and there is a continuum from one to the other but hippos tend towards the extreme K-strategist, much more so than their supposed close relatives, the pigs. From a conservation point of view, K-strategists present the greatest problems as they are less able than r-strategists to bounce back from a severe reduction in numbers. For this reason, it is hardly surprising that the species most at risk of extinction include a high proportion of large mammals. Hippos show some tendencies towards r-selection, however, such as the relatively short gestation period.

CHAPTER 6

DIET AND FEEDING HABITS
OF HIPPOS

Common hippo grazing

Both species of hippos are herbivorous but because of their very different habitats, they differ considerably in the food that they eat. Despite their watery environment, neither feeds on aquatic vegetation to any great extent.

It is not easy to find out what hippos eat, or what any animal eats, for that matter, but some techniques available for large mammals have been described by Field (1968a,b). An obvious way, at least in theory, is simply to see what the animal is taking into its mouth. In practice there are some problems. First, it is necessary to be close to the subject, easy enough with a domestic cow or horse but not so easy with a potentially dangerous animal like a hippo. One solution is to tame a hippo and follow it. This has been done but the method is not free of bias. Apart from anything else, the sample size, one, is small and what a single hippo eats may not be typical of the species. If one followed just one person and recorded food intake, one might suppose that the whole human race feeds on hamburgers or fish and chips, depending on the taste of the experimental subject. A further problem with the use of tame animals in feeding studies is that the subjects may have developed what veterinarians call "depraved tastes", such as a predilection for paper bags, particularly if, as is usual, they have not been fed on natural foods in captivity. Despite this, the method can be very useful as it is possible to see what wild foods the animal is rejecting as well as what it is selecting. This is not possible with most other techniques.

Another method for detecting food preferences is to examine the vegetation after the herbivores have passed through. It is usually possible to see which plants have had their leaves or shoots nipped off and which have not, although, unlike the previous technique, one cannot be certain that the animal is positively rejecting a plant rather than just not noticing it. The method is also limited to occasions when only one species of herbivore is involved.

One of the simplest ways of investigating foods is to examine dung, on the principle that what goes in must come out. This is true up to a point but it does not take differential digestion into account. Soft plants are likely to be more completely digested than those with a high lignin content so that remains of the latter will appear at a proportionately higher rate in the dung than in the diet. A further drawback is that the method is tedious and time-consuming to a degree if the fullest information is to be extracted from the dung sample. It is often easy enough to separate the remains into broad categories, such as leaves or stems, but if species need to be identified, it is necessary to carry out an elaborate treatment of the dung with chemicals to allow microscopic inspection of markings on the cuticle of the plants, as these are diagnostic of the species. This first requires a reference collection of plant cuticles to be assembled so that comparisons can be made with known species. The technique, pioneered by Stewart (1965), is one of the more effective methods for studying feeding habits although it has not been widely used, possibly because of its complexity.

The problem of bias from differential digestion in dung analyses can be considerably reduced by examining stomach contents as, normally, very little digestion will have taken place. This usually requires killing the animal and consequently it is not a popular technique for use on endangered species. If animals are being killed anyway for other reasons, it makes sense to take the opportunity to examine the stomach contents. This has been possible with the large hippo in several of the culling programmes that have taken place. Even though the plant remains in the stomach have not been so macerated as those in dung samples, it is still necessary in many cases to carry out preparatory treatment before the individual plant species can be identified from the markings on their cuticles.

A rough idea of the food of a herbivore can be obtained by recording the vegetation in regions frequented by a particular species. An animal that is consistently found in thick bush is likely to be a browser whereas one that is always in open grassland is probably a grazer. Another clue can often be obtained from the nature of the teeth but again, this usually takes one no further than a classification into grazer or browser. In the case of the common hippo, the large, flat molar teeth, specialised for grinding, tell us that it is primarily a grazer but we need to go into rather more detail than that. The gut anatomy also gives a clue to the type of food eaten as the lining of the stomach in ruminants differs between selective and roughage feeders, i.e. those that browse on tender shoots and those that eat coarse grass, and there is a variety of intermediate forms. How this varies in ruminants is discussed by Hofmann (1974).

Rumination is more than chewing the cud. The plant food that mammalian herbivores eat contains much structural material such as cellulose and lignin, neither of which can be digested by mammals, which lack the appropriate enzymes. Cellulose is an important energy source for herbivores but they have to rely on micro-organisms – protozoans and bacteria – to carry out the digestion on their behalf. The arrangement suits both parties – the micro-organisms are provided with an ample source of food and the herbivores absorb the monosaccarhide sugars produced by the micro-organisms through fermentation of the cellulose. The mammal also benefits by digesting the dead bacteria with their own gastric juices so gaining access to the volatile fatty acids in the bodies of the micro-organisms. These large herbivores, therefore, effectively synthesise their essential fatty acids, which other mammals have to take in with their food. The process occurs higher up in the gut in ruminants and so there is a greater length of small intestine available for the absorption of the products of digestion.

The fermentation chamber in the ruminant is the rumen, in which the food is stored while digestion takes place. After a preliminary breaking down of the ingested forage, pellets of the food are regurgitated for further chewing. This breaks down the food into smaller particles, which provide the maximum surface area for the bacteria to attack, so increasing the efficiency of digestion. The duration of rumination depends on the nature of the food. The process is prolonged with coarse dry grasses but much shorter with succulent shoots or leaves. Rumination is started by the production of stomach gases, such as hydrogen, carbon dioxide and methane, which cause the production of sugars. Excess gas is expelled through the nostrils, not the anus as is often erroneously stated, by a process known as eructation.

A ruminant has to be large in order to carry a large enough fermentation vat around with it. There are small ruminants but they usually feed on high quality browse such as buds or shoots and some are semi-carnivorous. Grazers are usually large since grass contains much structural material requiring prolonged digestion. Small grazers, such as the rabbit, cope by refection, in which soft pellets from the caecum are passed through the anus and recycled through the gut.

Ruminants are not the only large mammals that eat coarse vegetation, e.g. horses do, and one might wonder how they manage. They, too, rely on fermentative digestion but the fermentation chamber is the caecum and colon, not the stomach, which is simple and single-chambered, i.e. the animal is monogastric. It is also known as a hind-gut digester, to distinguish it from the ruminant fore-gut digester, but, as mentioned earlier, this is a misnomer as the caecum and colon are part of the embryonic mid-gut.

It is likely that rumination has evolved at least three times in the artiodactyls; namely in the camels, cervids (deer) and bovids (antelopes and cattle). Other ruminating groups that may have evolved the habit independently are giraffes and tragulids (mouse deer and chevrotains). It appears to be an efficient way of dealing with food that is difficult or impossible for a mammal to digest but it is not the only solution to the problem for there are plenty of apparently successful non-ruminants such as elephants and the perissodactyls (tapirs, rhinos and horses). The end product is the same but as the food has to pass some way down the alimentary canal before fermentation can occur, the process is generally considered, without much evidence, to be inferior to rumination. It is only when one looks at the relative number of species in each group that one begins to wonder if it is rumination that has led to the artiodactyls being such a vigorous, expanding group while the perissodactyls are in a sad decline. In the Pleistocene, there was a large number of perissodactyls but now they are limited to two tapirs, five rhinos and six or seven horses, all of which, with the exception of the plains zebra, are rare or in imminent danger of extinction.

Although it does not chew the cud, the hippo enjoys many of the benefits of rumination by having a chambered stomach as its fermentation vat. It would be expected, therefore, to feed on high quality food and this is indeed the case for the short, young grass on which it feeds is higher in proteins and soluble carbohydrates (sugars) than the coarser surrounding grasses.

In view of the radically different life styles of the two hippos, each species will be considered separately in the following account of their food and feeding behaviour. Most studies have been carried out on the large hippo and very little is known about the food of the pygmy hippo in the wild but its internal anatomy is not very different from that of the larger species.

THE DIET AND FEEDING HABITS OF
HIPPOPOTAMUS AMPHIBIUS

Mention has already been made of the coarse, dry nature of the gut contents, which are very different from that of ruminants. The dung is also coarse and more like that of horses than of ruminants. This suggests that digestion is less efficient than in a ruminant but there is evidence that the food is retained in the gut for a longer period to give more time for fermentation to occur. The evidence comes from the weight of food consumed, which is very low for an animal the size of a hippo. The mean dry weight of the stomach contents of hippos culled in Uganda was 34.9 kg for males and 37.7 for females, equivalent to only 0.95 to 1.3%, respectively, of the body weight. If the food passed through quickly, the percentages in terms of the total weight of the gut contents should be of the same order of magnitude but in fact, they are very much higher, averaging 21% in males and 22.8% in females (Arman & Field, 1973). These figures are very high compared with weights of the gut contents, which varied from 8.7 to 17.3%, in 14 ruminants examined by Ledger (1968).

These figures gave rise to the suspicion that hippos leave the water only every other night to feed. Examination of the stomach contents of culled hippos in Uganda (Field, 1970) lends support to this suggestion. Results suggest that digestion does not begin until noon and that the stomach contents are halved by about 18.00h. The inference is that the stomach fill represents two nights' feeding. The results of Pienaar *et al.* (1966) lend credence to this suggestion but there is no direct evidence that hippos go ashore on alternate nights. It is true that hippos can be found in the water during the night but it is known that some return for a rest in the middle of feeding (p. 53).

Feeding technique

The grazing technique of the hippo is unusual, although it is similar to that of the white rhinoceros. The hippo plucks the grass with its lips, not with its front teeth, which play no part in feeding. As it progresses across the grassland, the hippo swings its head from side to side with its muzzle near the ground at the lowest point of the swing. This is the moment when the hippo grabs the grass leaves with its wide, toughened lips. As the head is swung away, the leaves are nipped off and passed to the back teeth for mastication. Coarse, tussock-forming grass species are not suitable for this treatment as the stalks tend to slip between the lips and are not broken off. The technique works best with short, creeping species, which are lifted from the ground with the lower lip before being taken. This grazing habit results in a very short grass sward that would do credit to a golf course and the appellation of "hippo lawns" for the grazed regions is apt. It is unusual for such a bulky animal to feed on such short grass but the hippo is able to do so because of the anatomy of its mouth, which is notable for the breadth of its jaws. In addition to creating such lawns, the hippo maintains them by preferentially feeding on them. Grasses nearby grow to a greater height and become unattractive to hippos even though they are of the same species. Murray & Illius (1996) pointed out that some species can "capture" a sward for their own use by altering its physical structure or floral composition so making it suitable for the species in question but unattractive to others. The hippo is a very good example of such a species. The inverse relationship between the numbers of hippos and buffaloes on Mweya Peninsula in Uganda, mentioned in the next chapter, can be explained on this basis.

A typical hippo grazing ground, therefore, consists of a mosaic of closely cropped lawns interspersed with areas of long grass. It is only when hippo density rises above a certain level that the longer grass is exploited and, in turn, is reduced to a lawn-like state. This close grazing has significance in management as there is a propensity for large concentrations of hippos to cause soil erosion by reducing the protective grass cover and so exposing the soil to wind and rain. This aspect is discussed in some detail in Chapter 9.

The food of hippos

The vast majority of the food eaten by hippos is grass but some dicotyledons (broad-leaved plants) are bound to be ingested incidentally along with the grass. Hippos will occasionally take some aquatic vegetation. Field (1970) mentions that he has seen hippos eating the floating plants of the Nile cabbage (*Pistia stratiotes*) as I have myself, but they appeared to be merely toying with the plants in a rather bored, desultory way and it is unlikely that many were actually eaten. Mugangu & Hunter (1992) report instances of hippos in the Virunga National Park, Zaire, feeding more extensively on aquatic vegetation as a response to food shortage. They measured the crude protein contents of the aquatic as well as the terrestrial plants and concluded that the grasses, except for *Panicum repens*, were not providing enough protein to sustain the hippos so that the animals were obliged to turn to alternative foods. In explanation, they suggest that the hippo population had exceeded the carrying capacity of the park. Bere (1959) reports the results of an examination of 122 hippos shot in the Queen Elizabeth National Park, Uganda, during a control operation in 1958 and states that the stomach contents contained nothing but grass. The avoidance of aquatic plants is of significance in any discussion on the possible reasons why hippos have taken to water in the way that they have. Clearly it is not so that they can gain better access to food plants.

Fruit is a highly nutritious food but it is neglected by the large hippo although fruit-eating is found in many ungulates that share its range and in its close relative, the pygmy hippo. Some tree fruits are eaten – Dunham (1990) found that fruits of *Acacia albida* made up 1.9% of the faecal mass – but it is possible that the fruits, which are very small, could have been ingested accidentally.

There are not many studies of hippo feeding that have identified the individual species of grass consumed. One of the first, and still one of the best, is that of Field (1968a,b, 1970, 1972), who analysed the stomach contents of hippos shot in Uganda during the management culling carried out in the 1960s. The hippos were taken from four areas in the Queen Elizabeth National Park and samples of the stomach contents were collected from a minimum of six specimens from each area at four times of the year. Where possible, the six animals comprised two adult males, two adult females and two juveniles. The national park lies on the Equator and so experiences four seasons a year, two wet and two dry, each roughly of three months' duration. Hence a total of 24 hippos was examined from each area except for Lion Bay, on the Ankole shore of Lake Edward, where 126 hippos were collected, making a grand total of 198. As each sample took some two hours to analyse, it is easy to understand why the method has not been more widely used.

Field (1972) compared the percentage distribution of hippo stomachs that contained certain grass species with the percentage distribution of the grasses growing in the study area. Table 6.1 shows the results for ten grass species from one of the areas (Lion Bay) as a typical example. If the hippos were taking grasses in proportion to their abundance, the two distributions should be the same but it can be seen that they are not, suggesting that some species are selected and others avoided.

The percentage occurrence figures in Table 6.1 show the results from only one

Table 6.1 Comparison between the percentage occurrence of grass species in 24 stomachs of hippos and the percentage occurrence in grasslands. Data for Lion Bay in the Queen Elizabeth National Park, Uganda. The hippo data refer to the four seasons in the year. (After Field, 1972.)

Plant species	*% occurrence in hippo gut*				*% occurrence in grassland*
	wet	*dry*	*wet*	*dry*	
Grasses					
Bothriochloa spp	100.0	66.7	66.7	100.0	28.0
Brachiaria decumbens	66.7	83.3	83.3	100.0	8.0
Cynodon dactylon	33.3	83.3	100.0	100.0	28.0
Chloris gayana	100.0	100.0	100.0	100.0	36.0
Heteropogon contortus	83.3	66.7	16.7	33.3	6.0
Hyparrhenia filipendula	66.7	33.3	33.3	83.3	26.0
Panicum repens	0.0	50.0	16.7	0.0	locally common
Sporobolus homblei	0.0	0.0	0.0	16.7	very rare
Sporobolus pyramidalis	100.0	100.0	100.0	100.0	56.0
Themeda triandra	100.0	100.0	50.0	100.0	9.0
Other grass species	100.0	100.0	100.0	100.0	abundant
Dicotyledons	50.0	66.7	100.0	50.0	abundant

area. The grasses are not necessarily eaten to the same extent elsewhere but those that are consistently taken in all areas, whether abundant or not, comprise *Bothriochloa* spp, *Chloris gayana*, *Heteropogon contortus* and *Sporobolus pyramidalis*. Those eaten in most areas include *Brachiaria decumbens*, *Hyparrhenia filipendula*, *Themeda triandra* and *Cynodon dactylon*. Those eaten in only a few areas are *Panicum repens* and *Sporobolus homblei*.

The analyses so far tell us only which plants are eaten but do not provide information on the quantity of each that is consumed and for this, the actual numbers of plant particles need to be counted. The basic data can be found in Field's publications, but the means for the number of fragments belonging to each grass species expressed as percentages of the total number from all species are included in Table 6.2. This quantitative approach is far more meaningful and some conclusions of great interest can be drawn.

It can be seen that only three of the ten grass species examined, were present in the stomachs at densities that, averaged over the four study areas, exceeded 10% in one or both seasons. These were *Brachiaria decumbens*, *Cynodon dactylon* and *Sporobolus pyramidalis*. Although the figures are not shown here, there were, nevertheless, localities where the selection for certain species was much higher than this and only two, *Hyparrhenia filipendula* and *Sporobolus homblei*, never exceeded 10% in any area or season although *Panicum repens* came close to joining them. Dicotyledons were insignificant except possibly in one area in one season when the amount in the stomachs rose to 7.3%. This finding supports the assumption that such plants are taken accidentally.

Hippos, therefore, appear not to take food plants in proportion to their occurrence but to exercise selection and rejection although the latter need not be absolute. We all

Table 6.2   The average percentages of fragments from each plant species in the stomachs of hippos from four study areas in the Queen Elizabeth National Park, Uganda in wet and dry seasons of 1962-1963. (After Field, 1970.)

Plant species	Mean % of fragments	
	Dry	Wet
Grasses		
Bothriochloa spp	4.2	6.3
Brachiaria decumbens	16.3	9.4
Cynodon dactylon	8.2	13.6
Chloris gayana	7.0	7.2
Heteropogon contortus	8.3	
Hyparrhenia filipendula	4.0	
Panicum repens	1.4	2.1
Sporobolus homblei	1.5	1.9
Sporobolus pyramidalis	24.7	22.3
Themeda triandra	9.9	9.1
Other grass species	2.9	4.9
Unidentified grasses	8.7	9.7
Dicotyledons	3.0	1.2

have preferences and although we might prefer fillet steak to scrag end of neck, we would accept the latter if nothing else were forthcoming. Species consistently present in the stomachs of hippos but only in small quantities, despite being abundant in the pasture, were *Bothriochloa* spp, *Chloris gayana* and, with some exceptions, *Hyparrhenia filipendula*. Such grasses may be said to be avoided. Species present in the stomachs in large quantities and also abundant in the pasture were *Heteropogon contortus, Sporobolus pyramidalis, Cynodon dactylon* and *Themeda triandra* although, again, there were some exceptions in certain areas. These species were presumably consumed in proportion to their abundance. Finally, there was one grass species, *Brachiaria decumbens,* that was common in the stomachs in most areas but rare in the pasture. Such a grass can be said to be preferentially selected.

Tables 6.1 and 6.2 are also of interest in showing that there were marked seasonal variations in the selection of grass species by hippos. An inspection of Field's paper shows that *Bothriochloa* spp, *Cynodon dactylon* and *Panicum repens* constituted a greater proportion of fragments in the wet season. Other unidentified grasses, mainly annuals, were also more common in the rains. *Brachiaria decumbens* fragments were more frequent in the dry season as were those of dicotyledons.

The finding that hippos can be selective in the grass species eaten raises the question of how they do it. Selective feeders need to manipulate food items. Primates, which have hands, and antelopes, which have mobile lips, are able to handle small objects and can strip leaf from stem, for example. Such dexterity is beyond a hippo with its wide mouth and "blind" method of plucking grasses. One likely explanation is that they do not select for grass species as such but merely select the patches of sward in which to feed. Such swards often consist of only one or a few grass species and in selecting where to feed, the hippo is selecting what species to feed on. The hippo, itself, helps to create its own choice feeding stations because of its close grazing. Such a habit encourages the growth of the short, creeping grasses that the hippo favours. Other, less palatable, grasses are shouldered out by the more aggressive species.

Hippos are not the only grazers in the African rangelands and an obvious question is how do they interact with other herbivores? Do they compete or is there ecological separation between them? In theory, one would expect them to have evolved in such a way as to refrain from competition on the grounds that it is better to avoid confrontation than to get into a fight you might lose. Competition can arise in a number of ways but in feeding, it usually results from two species seeking the same food item, such as a grass species. In such circumstances, if one species took only the leaf and the other only the stem or inflorescence, competition would be avoided. This is termed ecological separation or resource partitioning and there is sound evidence that such a process has occurred many times. However, if one species were much more efficient at foraging, the second species would be unable to compete and would be eliminated. This is called competitive exclusion and again it is a well known phenomenon, which can be documented when an aggressive alien species is introduced into an area as with mammalian carnivores into New Zealand. In evolutionary terms, ecological separation can be looked upon as the Ghost of Competition Past.

Field (1972) considered the question of competition between the hippo and other herbivores. Competition exists if two animals, whether of the same or different species, exploit a resource that is in short supply. The resource in question may be an oestrous female or territory, in the case of intraspecific competition, or food or shelter in the case of interspecific competition. In the context of this chapter, we are interested in possible competition for food. The hippo's habit of close grazing is similar to that of the warthog and if competition were to occur, it might be expected to be between these two species. Direct, or interference, competition is unlikely, for the hippo is a nocturnal grazer whereas the warthog feeds by day. Hence there is no question of the hippo's chasing the warthog away from the pastures. Nevertheless, there could be exploitative competition and the warthog could outcompete the hippo if it removed all of the food during the day leaving nothing for the hippo. Note that competition can occur only if the resource is in short supply.

In all ecosystems the risk of competition over food is likely to be greatest during the period of deprivation. In the case of herbivores, this is the season of lowest plant production, i.e. winter in higher latitudes or the dry season in the tropics. In the wet season, there is a superabundance of plant production resulting in a surplus of food for herbivores.

Field (1972) found that there was little evidence of competition for food in the wet season between any of the species he investigated. In the dry season, on the other hand, the degree of overlap between food plants was sufficient for the possibility of competition to arise. Species with which the hippo might find itself in competition were buffalo, warthog and waterbuck. Competition with waterbuck for *Sporobolus* was possible in two of the study areas and for *Heteropogon* in a third. Both buffalo and warthog could potentially compete with hippos for *Sporobolus* in a fourth study area. This is not to say that competition did occur but that it was possible. Much would depend on the amount of grass available and its nutritional quality.

Far from competing, many herbivores show facilitation whereby the presence of one species enhances the welfare of another (Gwynne & Bell, 1968). This appears to be so with hippos and buffaloes (Field, 1968c). *Sporobolus* grassland is utilised by buffalo but its presence is a consequence of hippo grazing. The buffaloes repay the favour by feeding on the tussock grasses, so keeping the sward open and allowing the creeping grasses, preferred by hippos, to spread. This may be true only up to a point for when hippo numbers increase beyond a certain level, buffalo numbers decline (p. 98).

A rather one-sided facilitation between hares and hippos was suggested by Ogen-Odoi & Dilworth (1987). They found the density of hares to be much higher in a short grass study area in Queen Elizabeth National Park than in other habitats and suggested that the grazing activity of hippos was responsible. They also suggested that the presence of the hippos frightened away potential predators of the hares.

Carnivory in hippos

It may come as a surprise to hear that hippos have been seen feeding on meat on a number of occasions and even indulging in cannibalism. The first reported cases was recorded in Hwange National Park, Zimbabwe, in 1995 by Dudley (1996) when the species eaten was impala. One instance featured predation as well as scavenging but another involved only scavenging. The predation occurred when a male impala, attempting to escape from a pursuing wild dog, swam through a group of 18 hippos resting in Masuma Dam in the north of Hwange National Park. The hippos killed the impala and proceeded to eat it. Within the hour, five hippos broke away and attempted to scavenge another impala that had been killed at the water's edge by the wild dogs. Despite determined efforts by the hippos, the dogs managed to consume most of the carcase themselves. The second instance reported by Dudley occurred two months later and again concerned an impala, which had been killed by two crocodiles at Nyamandlovu Pan in the east central region of Hwange National Park. A subadult hippo was seen to join the crocodiles and eat a portion of the carcase.

The incident in Masuma Dam is ironical as a report of a hippo saving an impala from wild dogs in the same Dam is given by Erwee (1996) in the BBC Wildlife magazine. In this case the hippo "guided" an exhausted male impala swimming in the Dam away from a pack of wild dogs, which had chased it into the water. The hippo then nudged the impala onto dry land and engulfed its body in its mouth several times without harming it "almost as if to breathe new strength into the impala", as the author rather teleologically put it. It was more likely a case of attempted predation but whatever the motive, the impala survived. As a footnote to this story, the editor of the magazine mentioned a similar case in the Kruger National Park in South Africa when a hippo tried to save an impala from a crocodile.

Two other cases of meat-eating have been reported from East Africa by R.D. Estes and involved wildebeest.

An instance of cannibalism was reported to me by Miss Nicky Dunnington-Jefferson, who has given me permission to describe the incident. She and a companion were relaxing one morning in Mvuu Camp on the banks of the Shire River in Liwonde National Park, Malawi, when they noticed a commotion on the other side of the river, which proved to be caused by a group of crocodiles feeding on a hippo that had died from fight wounds the previous night. The carcase had been towed across the river by the authorities to prevent any smell reaching the camp. Miss Dunnington-Jefferson and her companion noticed a hippo amongst the crocodiles and, in a boat provided by the Manager, they approached more closely to find that the hippo was biting chunks off the carcase and eating them. They filmed the scene and kindly sent me a copy of the video, which clearly shows the hippo eating the flesh of its erstwhile companion. Still photographs of the scene are reproduced in Plates 1 & 2.

These reports from widely separated localities suggest that carnivory is a feature of hippo biology, if somewhat rare. The killing of other animals by hippos is not surprising and has often been witnessed. Hippos can be aggressive animals and one that has suffered injury in a fight is especially likely to vent its feelings on the nearest animal to cross its path. Hence reports of "predation" by hippos should be treated with caution since predation implies the killing of an animal with the intention of eating it. The accounts given above cannot unequivocally be ascribed to predation but what is undoubtedly correct is the existence of scavenging by hippos. There are various possible explanations for this behaviour.

The first is that the behaviour is aberrant and confined to the hippo equivalent of psychopaths. Alternatively, the carnivory may be fulfilling a nutritional need. Vegetation may lack essential nutrients or trace elements necessary for a healthy metabolism but such elements are always present in meat. A lack of phosphorous may lead to herbivores gnawing bones. This is not often observed but Wyatt (1971) witnessed two giraffes on separate occasions chewing bones of unidentified bovids in Nairobi National Park, Kenya. The widespread habit of herbivores consuming soil in order to obtain "salt" is also a reaction to a mineral deficiency in the diet. It is significant that the hippos feeding on meat in Zimbabwe had previously been observed eating elephant dung and twigs, suggesting that something was missing from their normal diet. This dietary hypothesis is probably the correct explanation for carnivory in hippos. The alternative – that hippos are essentially omnivorous – is unlikely and meat eating should be regarded either as aberrant behaviour or as a reaction to nutritional stress. Certainly, the gut anatomy is unsuited to a carnivorous diet.

The Diet and Feeding Habits of *Hexaprotodon liberiensis*

So far as I know, there has been no study of the feeding habits of the pygmy hippo comparable to those into the habits of the larger species apart from the study by Hentschel (1990) in the Ivory Coast. Most other information comes from secondary

TABLE 6.3 The most important food plants taken by pygmy hippos in the Ivory Coast. + plant detected, − plant not detected. (From Hentschel, 1990.)

	Method of detection		
Species	*Dung analysis*	*Indirect methods*	*Direct methods*
FERNS			
Nephrolepis biserrata	−	+	+
Pteris burtonii	−	+	−
Athyrium proliferatum	−	+	−
Ctenitis variabilis	−	+	−
DICOTYLEDONS (Broad-leaved plants)			
Desmodium ascendens	−	+	−
Geophila afzelii	+	+	−
Geophila hirsuta	−	+	−
Neuropeltis acuminata	−	+	−
Ipomoea batatas	+	+	−
Justicia tenella	−	+	−
Dissotis rotundifolia	−	+	+
Physacanthus nematosiphon	+	+	+
Staurgyne paludosa	+	+	+
MONOCOTYLEDONS (Grasses etc.)			
Crytosperma senegalense	−	+	+
Floscopa grandiflora	+	+	+
Maschalocaephalus dinklagei	+	+	−
FRUITS			
Parinari chyrsophylla	−	+	+
Berlinea grandiflora	+	+	+
Anthonotha fragrans	−	+	−
Cuervea macrophylla	−	+	+

sources, although the observations are presumably based on firsthand experience of the animal. Most authors emphasise the more terrestrial nature of the pygmy hippo so it too does not feed on aquatic vegetation. Nor does it feed on grass to any extent for there is very little in the forests that it inhabits, except along the river banks or in glades. Where grass occurs it is readily eaten. Various forbs growing on the forest floor, as well as sedges, constitute an important part of the diet, as do fallen fruits. Leaves of bushes and trees are also eaten if they are within reach. Some plants mentioned by Hentschel are taken only up to 10 or 25 cm above ground level. The pygmy hippo feeds in the same way as its larger relative by nipping off the vegetation between its thick lips.

Some details of plants eaten by pygmy hippos in the Ivory Coast are given by Hentschel whose techniques varied from indirect methods, such as noting evidence of feeding on plants, to feeding experiments. Numerous lists of plants eaten by the pygmy hippo are given from which the most important of each type (those allotted four stars on a quantitative basis) have been extracted and are presented in Table 6.3. The techniques used to identify the presence of the plants are included. A plus sign means that the plant was detected by the method in question in one or other study area (Tai or Azagny) and a minus sign means that it was not detected.

This analysis suggests that indirect methods are more effective than the others in identifying food plants. It is usually very obvious that a plant has been fed upon by a herbivore but it is not so easy to identify the animal responsible. This is less of a problem in forests than on the open savanna as there are fewer potential browsers or grazers. Even if not seen the herbivore responsible can usually be identified from nearby tracks.

Altogether 83 plants comprising 17 ferns, 26 dicotyledons, 16 monocotyledons and 24 fruits are listed by Hentschel as food plants. The lists are mutually exclusive, i.e. no plants were recorded as losing both leaves and fruits to hippos. The prominence of ferns is perhaps surprising but these observations confirm the importance of fruits in the diet of the pygmy hippo and the wide range of species taken suggest that the pygmy hippo will eat whatever plants are available.

CHAPTER 7

THE ECOLOGY OF HIPPOS

'Muck-spreading' by a common hippo

EFFECTS OF HIPPOS ON THE ENVIRONMENT

This chapter is concerned with *Hippopotamus amphibius*, the larger of the two species. Hippos are big animals and, like other megaherbivores, they can have a profound effect on their environment, both biotic and abiotic. Their heaviness and relatively small feet mean that the pressure on the ground exerted by each foot is very high. This results in compaction of the soil so that when it rains, the water is likely to run off along the surface rather than soak into the ground. Apart from the risk of sheet erosion that might ensue, the run-off deprives the soil of the beneficial effects of rainfall. The hippo's practice of cropping the grass close to the ground adds to the danger of erosion as it leaves no vegetative cushion to absorb the force of heavy rain. Hippos can also be responsible for gully erosion. When they leave the water to feed, they tend to follow set trails, and the repeated wear of hippo feet may lead to the formation of a shallow trench. If there are a lot of hippos, the trench is not all that shallow and can become a gully, particularly if there is a sharply rising bank, in which case the gully becomes a miniature river when it rains. In time the gully may develop to a depth of 20 m or more and increase in length as it erodes farther back into the rangeland. The top soil that is washed away finishes up in the river or lake where it may form an alluvial fan if the water is sufficiently still. In marshy conditions, the hippo trails serve an important function as drainage channels.

This is particularly true of the Okavango Delta in northern Botswana, which has recently been studied in detail by McCarthy *et al.* (1998). The Okavango River is unusual in that it does not reach the sea but instead soaks into the sands of the Kalahari Desert. The river enters the delta as a "panhandle" along a series of channels up to 100 m in width and with a depth of 3–4 m. The slope is not steep because the river flows from an elevation of about 1000 m as it enters the panhandle to 975 m at the Gomare Fault, about 110 km downstream, where it spreads into the alluvial fan of the delta. The gradient remains shallow with a drop of only some 40 m over the remaining 180 km of the delta's length. Permanent swamps develop along the main channels and contain large lakes resulting from ancient oxbows. Seasonal swamps are formed during the rainy season, more than doubling the area of swampland. Low islands, rarely more than 1.5 m high, occur within the swamps and vary in size from the 80×15 km Chief's Island to individual termite mounds.

Large channels drain the permanent swamps in the north-east, becoming narrower and shallower towards the central regions. These channels are not permanent and, over time, the water leaks away through the peaty substratum, sediment is deposited and the channel bed rises above the level of the surrounding land. Water flow then slackens. This causes the water level to rise leading to an even slower flow and the channel becomes overgrown with aquatic plants. Eventually the channel dries out and after the peat is burnt in seasonal fires, the area reverts to dry land or seasonal swamp. Meanwhile, fresh channels are formed elsewhere.

These are changes due to geology and geography but the hippo plays an important rôle in the dynamics of the system by acting as a catalyst. Although they occur in the panhandle, hippos do not have much influence on the local topography but they do have an effect in the delta proper. Hippos there are abundant in the permanent swamps and use the islands as well as the surrounding grasslands as grazing areas. In moving from water to land they make use of the main channels, which run along the length of the delta, but they also create subsidiary channels at right angles to them. Hippos continue to use the channels while they are drying out and, by walking along the stream bed, keep the channels clear of encroaching vegetation. This undoubtedly prolongs the life of the channels but the hippos are unable to prevent them from ultimately drying out.

The hippos create a network of small channels between their daytime retreats and grazing grounds. These are important in the hydrology of the delta in that they lead to the development of new channels when previous ones disappear. The dying channel discharges large quantities of water into the back swamps but the hippo trails act as filters and direct the flow into suitably oriented small channels, which, with the increased flow, soon develop into full-sized ones.

The side trails caused by the hippos as they leave the main channels to reach the grazing grounds also have a significant effect on the system. Water gradients along these tributaries may be steeper than along the main channels and result in the transport of material from the bed of the latter. This builds up the level of the side channels and when they too degenerate into dry land and are burnt off, they remain as sandy ridges more than a metre high and may later develop into islands.

The topography of the delta, therefore, owes much to the activities of hippos and their removal would result in wide-ranging changes in the Okavango Delta. This is a cogent argument for the conservation of the species in view of the importance of the delta to other wildlife as well as its value as a tourist attraction. These hydrological effects are probably to be found in many, unstudied swamplands containing hippos.

In addition to the effects on its inanimate environment, the hippo influences other living forms sharing its range. The most obvious effect is on the vegetation. The hippo's feeding technique was described in the last chapter, where it was pointed out that the consequence is a sward of short creeping grasses. As is so often the case with large mammals, an equilibrium is set up between a high herbivore biomass and a low plant biomass. The equilibrium is stable and if one or other of the factors is altered slightly, the system reacts to restore the *status quo ante*. The stability acts only within limits and if there is a massive perturbation, the system may switch to a new equilibrium, which could be unstable. It cannot remain unstable for long and eventually either a new stable equilibrium will be reached or the system will crash. Not all ecologists are convinced that a stable equilibrium is ever established in a natural ecosystem and it must be admitted that most studies of the population dynamics of large mammals rely on computer simulation based on a minimum of field observations, but the wildlife manager, who has to take the decisions, cannot afford the luxury of hiding behind academic debate. It is safer to prepare for the worst-case scenario but at the same time not to take measures that would be irreversible should the management action prove to be incorrect.

The high herbivore/low plant biomass equilibrium is potentially more risky than the opposite situation since the biomass of the vegetation is closer to zero and vulnerable to a sudden increase in herbivore numbers, which would lead to overgrazing. If this were to happen, the system should in theory restore the balance through an increased death rate and/or a reduced birth rate in the herbivore. With large herbivores, however, there are problems. To begin with, they are starvation-tolerant and can survive on an inadequate diet for far longer than a small herbivore, such as a mouse or vole, could. Hence they do not die at once but survive to cause further damage to the vegetation so that the carrying capacity continues to fall. Another problem is the relatively low reproductive rate of these *K*-selected animals, which makes it difficult for them to regulate their population size quickly enough to keep pace with rapid, human-induced changes, such as a greatly increased mortality caused by poaching. Finally there is their longevity. Even if they stopped breeding altogether, their long life span would ensure that plenty of animals were left to overgraze. In nature, such a system would be unsustainable and the herbivore population would either die out or, more likely, move elsewhere to an undamaged section of their habitat.

The repression of grasses by close hippo grazing has far-reaching effects on other types of vegetation. Without a grass cover, grass fires are obviously impossible and as fires tend to prune bushes by killing young side shoots, the absence of fire leads to a spread of thickets. Thus there is a replacement of grassland by bushland so that the hippos, through overgrazing, destroy their own habitat as well as that of other grazers. Thus in addition to influencing the species composition of the plants, they also change

Plate 1 Hippo and crocodiles feeding on the carcase of a hippo at Mvuu Camp on the Shire River in Liwonde National Park, Malawi on 1st December 1996. The dead hippo had succumbed after fighting and had been towed across the river to reduce the smell in camp. It was also decapitated. *Photo: Miss Nicky Dunnington-Jefferson.*

Plate 2 Hippo tearing at the flesh of the dead hippo. Substantial quantities of meat were consumed. *Photo: Miss Nicky Dunnington-Jefferson.*

Plate 3 A female hippo and her well-grown calf leave the water by day to loaf in the sun. Hippos never move far from water in the daytime and will slip back in if they start to over-heat. *Photo: S.K. Eltringham.*

Plate 4 A school of hippos in a permanent river. Such a school comprises females and calves for the most part, under the control of a territorial bull although some non-breeding males may be among them. *Photo: S.K. Eltringham.*

Plate 5 A group of mostly male hippos towards the end of the dry season in a temporary wallow that is beginning to dry out. Water is not necessary for a hippo's well-being as long as the skin is kept wet, and mud serves well enough for that. The presence of temporary wallows in the wet season means that hippos are able to extend their feeding range without expending excessive energy in walking to grazing grounds distant from permanent water. *Photo: S.K. Eltringham.*

Plate 6 The result of over-stocking with hippos. Almost all grass has been removed by grazing pressure and bushes have begun to encroach onto former grassland. The soil has also been comapacted by the feet of the heavy animals so that rain does not percolate through the soil but runs off, causing sheet erosion. *Photo: S.K. Eltringham.*

Plate 7 A badly eroded lake shore as a result of hippo grazing. There is extensive sheet erosion and the beginnings of gully erosion. On steeper land, the gullies would be much deeper. As it is, there has been considerable loss of top soil as indicated by the "cliffs", whose tops reveal the original level of the ground. *Photo: S.K. Eltringham.*

Plate 8 A lake shore in Lion Bay, Queen Elizabeth National Park, Uganda, a few years after the removal of hippos by culling. When hippos were present, the vegetation was in a similar condition to that shown in Plate 7. *Photo: S.K. Eltringham.*

the make-up of the animal communities.

These aspects are discussed in some detail by Laws (1968b), who emphasises the keystone function of the hippo in the tropical ecosystem. A keystone species is one that plays a pivotal role in community ecology such that changes in its numbers can have far-reaching, knock-on effects on other species, both plant and animal. The other keystone species considered by Laws is the elephant. This species tends to convert woodland into grassland through its destructive effects on trees and the hippo helps to maintain grassland through its grazing activities.

THE HIPPO PROBLEM IN UGANDA

These possibilities are not purely hypothetical for such a situation is believed to have arisen between hippos and grasslands in western Uganda during the 1950s (Bere, 1959). The saga is worth telling in some detail as it was probably the first large-scale management project in an African national park and it demonstrates the importance of scientific research when applied to wildlife management.

The Queen Elizabeth National Park was established in 1952. It includes parts of two lakes, Edward and George, which are very productive of fish, and several fishing villages have grown up on their shores. The fishing industry was too important to suppress and so when the park was gazetted, the villages were allowed to remain physically within the park although legally they were excised from it. It is not known how much hunting of hippos occurred previously but it may well have been high enough to keep hippo numbers within bounds. With the control of poaching that came with the park, an important mortality factor may have been reduced to a sufficient extent to have allowed the hippo population to increase to a level whereby habitat damage occurred. We shall probably never know but, whatever the reason, the grasslands were being replaced by bare soil over which dense thickets were spreading (Plate 6). During the 1950s this did not appear to worry the park managers, who probably felt that any increase in a favoured species was a good thing. It was not long, however, before a warning note was sounded by some American biologists who had arrived to study the animals in the park as part of a scheme to repay Britain's war debts.

During the Second World War, the United States provided Britain with material aid on a lease-lend arrangement. One of the ways in which this debt was repaid was through the British Government providing funds for American academics to carry out research in British institutions including those in the then British colonies and protectorates. The scheme was the imaginative brain-child of Senator J. William Fulbright, a Rhodes Scholar and erstwhile President of the University of Arkansas, so the fortunate academics were called Fulbright Scholars. Some of these were wildlife biologists and five of them were sent to Uganda. The first two, George Petrides and Wendell Swank, were consigned to Queen Elizabeth National Park and arrived in 1956. This was at a time when there had been very little research into large African mammals so they had no body of knowledge to fall back on in interpreting their observations. Their remit was to survey the large mammals and so they instituted a series of game counts in the course of which they could not help but notice the large number of hippos in the park. They also noticed the low grass cover and, putting two and two together, concluded that the hippos were destroying their own habitat. The solution, they suggested, was to reduce the number of hippos drastically by shooting some of them (Petrides & Swank, 1965).

At that time, the prevailing philosophy was that national parks were inviolate sanctuaries and the concept of wildlife management extended to little more than controlling predators. The suggestion that thousands of hippos should be shot in a national park was looked upon by many as equivalent to desecrating a cathedral in a particularly unpleasant way. Nevertheless, the proposal had been made and it was up to the national parks authorities to decide whether or not to act upon it. Wisely, they took soundings with many conservation bodies and individuals and although many misgivings were expressed, the general feeling was that the parks should grasp the nettle and go ahead with the cull. Alternative possibilities were tried including the draining and fencing of wallows to deter hippos from entering sensitive areas but the maintenance costs would have been too high, apart from which the wallows were important to other species for drinking and bathing. Culling, therefore, seemed to be the best solution and it is to the credit of the authorities that they agreed to it, given the almost revolutionary nature of the action. It was probably not so revolutionary to Americans, with hunting more deeply embedded in their culture, and no doubt the differing philosophy towards wildlife management in the United States was instrumental in the proposal being made in the first place. British biologists may well not have considered it to be an option.

By this time, Petrides and Swank had returned to the United States and it was left to a third Fulbright Scholar, Bill Longhurst, to advise the parks on what and where to cull. In order to make an inroad into numbers, it was agreed that the cull had to be on a massive scale. The taking out of only a few animals would not have had much effect and might even have stimulated a higher reproductive rate. Consequently the removal of half the population was proposed. Surveys made in the park had led to the belief that there were about 14 000 hippos altogether so a cull of 7000 was instituted. Counts made after the cull was concluded returned totals of around 14 000, suggesting that the original population was nearer to 21 000 and that a third, rather than a half, of the hippos had been removed. Details of the culling techniques are given in Chapter 9.

The effects of the reduction in numbers on the hippos

In addition to reducing numbers, the culling of hippos should have had an effect on their population dynamics. The usual reaction to an increase in the death rate of an animal population is a corresponding increase in recruitment. The latter can be achieved in a number of ways including a reduction in the age of sexual maturity and an increased calving rate. Such changes seem to have occurred in the Ugandan hippos but they were not clear-cut because reproductive increase was masked to some extent by immigration from unaffected sub-populations. Nevertheless, Laws (1968b) detected a significant increase in the percentage of calves from 5.9% before culling to 7.2% in 1962 and eventually to 14.0% in 1966. Part of this increase was due to the selective removal of older animals but an increased birth rate and a reduced calf mortality were also responsible. Pregnancy rates increased from 26% in 1958 to 33% in 1966, although the increase was not statistically significant. The age of puberty, based on ovarian evidence, fell from 12.1 years in 1958 to 9.7 years in 1962. All of these changes would have resulted in increased recruitment, in line with what would be expected of an animal population experiencing a massive perturbation in its numbers. Such a reaction is a healthy one for it means that the over-large population, which was presumably "trying" to reduce its numbers and hence was in a state of decline, is now in an expanding mode.

The effects of hippo removal on the vegetation

Before, during and after the hippo cull, surveys were made of the vegetation so that any changes coincidental with the reduction of hippos could be noted. The results of this research have been reported by Thornton (1971) for his study site on Mweya Peninsula. This is an area of about 4.5 km², which is almost surrounded by water and which lies at two levels, a low-lying, heavily bushed northern region situated some 30–37 m below a more open plateau. Deep gullies had formed on the steep, southern slopes and were spreading inland. These were said to be a result of hippo grazing although Bishop (1962) considered that they were a natural corollary of steep slopes and soft, unconsolidated sediments. Nevertheless, he thought that hippo grazing may have accelerated the spread even though it was not the principal cause. The dominant grass species was *Chrysochloa orientalis* with a 37.7% basal grass cover in May 1958, when Thornton began his surveys. He laid down 20 permanent transects, ten on the flats and ten on the plateau, each 100 ft (30.46 m) in length. These were surveyed at intervals over the next four years, although some of the transects could not always be relocated. The culling of hippos on the Peninsula did not begin until May 1958 so that the first survey made by Thornton represents the condition of the grasslands at the maximum grazing level. The results of these surveys are shown in Table 7.1.

The results were not all that might have been expected for there was no apparent immediate flush of grass. During the first year the percentage of grass cover decreased by about two-thirds from 14.7% in May 1958 to 5.4% in May 1959 but the amount of litter increased to 10.5%. As the litter is predominantly derived from grass, the combined cover for these two factors remained constant. The forbs (dicotyledons),

Table 7.1 Changes in percentage cover of vegetation and bare ground on Mweya Peninsula following removal of hippos in 1958. Figures are means for 20 transects except for 1960 and 1962 when only 17 and 15 transects, respectively, could be relocated. The annual rainfall is included for each year. (After Thornton, 1971.)

	May 1958	Oct. 1958	Feb. 1959	May 1959	May 1960	May/Jun. 1962
Grasses	14.7	8.5	5.4	5.3	9.1	10.9
Litter	1.2	8.3	11.6	10.5	19.5	27.2
Sedges	0.1	0.1	0.04	0.03	0.04	0.03
Dicotyledons	5.7	0.8	0.4	0.5	0.8	1.3
Bare ground	78.3	82.2	82.7	83.6	70.6	60.6
Annual rainfall (mm)	691.3 (1958)		675.1 (1959)		741.9 (1960)	560.4 (1962)

however, declined markedly from 5.7% to 0.5%. This decline was responsible for the increase in bare ground.

Grass cover had greatly increased by 1960 and two years later had almost returned to the pre-culling level. Litter also continued to increase and by 1962, grass and litter combined (i.e. grass production) represented a 28.1% cover compared with 15.9% four years earlier. As a consequence the proportion of bare ground fell considerably from a high of 83.6% in 1959 to 60.6% in 1962.

In summary, there was a general improvement in the rangeland with increased grass production and a reduction in bare ground. Forbs more or less disappeared and sedges, which were never common, declined almost to nothing. The recovery of the sward may not have been due entirely to the reduction of hippos because there was a distinct improvement in the rainfall between 1959 and 1960. The 1960 figure was considerably above the 15-year average of 714.0 mm for 1954–68. The rainfall between April and August, the long wet season at Mweya, is the most significant as it is the period when grass growth is at a maximum. In 1958 the rainfall was well above average in this season, which was also the period when the first batch of hippos were shot. The grasslands, therefore received the double bonus of a greatly reduced grazing pressure and an improved rainfall.

Changes in the individual grass species are shown in Table 7.2. At the beginning of the study, before any hippos were removed, the dominant grass was *Chrysochloa orientalis*, a stoloniferous perennial present in every transect and representing nearly 40% of the total grass cover. The next most common grass was *Cynodon dactylon* trailing somewhat behind with less than 10% cover and occurring in only a quarter of the

Table 7.2   Changes in the percentage grass cover for some of the more common grass species following removal of hippos from Mweya Peninsula. Details as in Table 7.1. Numbers in parentheses are the percentage frequency occurrence of the species on the transects. (After Thornton, 1971.)

Grass Species	May 1958	Oct 1958	Feb 1959	May 1959	May 1960	May/Jun. 1962
Cenchrus ciliaris	4.8	3.6	4.0	2.7	3.0	4.3
	(80)	(60)	(70)	(65)	(65)	(80)
Chloris gayana	6.4	6.0	16.3	6.8	7.2	19.4
	(30)	(50)	(55)	(40)	(76)	(80)
Chrysochloa orientalis	37.7	35.4	22.9	21.1	22.9	6.6
	(100)	(100)	(100)	(90)	(94)	(87)
Cynodon dactylon	9.2	8.4	1.4	2.1	3.2	11.7
	(25)	(35)	(15)	(25)	(35)	(60)
Sporobolus rangei	4.0	3.9	7.6	11.5	5.4	6.0
	(10)	(10)	(10)	(10)	(10)	(13)
Sporobolus pyramidalis	8.0	12.6	20.4	25.2	18.9	27.6
	(60)	(60)	(85)	(85)	(82)	(100)
Sporobolus consimilis	4.2	2.6	4.1	5.5	2.9	1.7
	(10)	(10)	(10)	(10)	(12)	(13)
Sporobolus stapfianus	7.0	10.3	9.7	10.7	7.6	3.1
	(40)	(35)	(40)	(45)	(35)	(47)

transects. *Sporobolus* species collectively accounted for over a quarter of the grass cover, with *S. pyramidalis* the commonest species.

Changes in the species composition of the sward soon became apparent with the reduction in grazing pressure. *Chrysochloa* showed a small reduction at first but went into a steep decline after a year, finishing with only a 6.6% cover in 1962 and no longer being recorded on all transects. The second most common species in 1958, *Cynodon dactylon*, showed a sharp fall at first but rallied to 11.7% cover at the end, higher than the 9.2% cover at the beginning. It was present on only a few of the transects, however, and too much should not be read into the fluctuations. It is sensitive to nitrogen and its decline was most noticeable near the lake shore, where there was a reduction in hippo dung after the culling. Species that showed increased grass cover included *Chloris gayana* and, particularly, *Sporobolus pyramidalis*. They were also present in most or all of the transects. These two species, with *Cynodon dactylon*, together covered 58.2% of the ground by the end of the survey.

The significance of the change in species composition can best be appreciated when the plant form is considered. *Chrysochloa orientalis*, the original dominant grass, is a creeping, low-growing plant of a type that thrives under heavy grazing close to the ground, such as it would experience from hippos. It has a low biomass but a high basal cover. Grasses that replaced it tended to be tufted or tall-growing and although they might not cover as large an area of ground, they were more significant in terms of aerial biomass, which Thornton did not measure. It is this fact that makes the percentage of bare ground, shown in Table 7.1 somewhat misleading. An increase in plant biomass may result in an increase in bare ground if tall leafy grasses replace short creeping species, which was the case. Under such circumstances the amount of litter would increase, as, indeed, happened. An increase in litter is beneficial as it is important in restoring degraded grassland. Thus, it reduces erosion by breaking the force of heavy rain storms and acts as a mulch in retaining water which would otherwise run off.

Three large gullies on the southern shore of the peninsula were monitored by Thornton by pegging. Between May 1958, when the pegs were inserted, and 1962 the gullies had advanced inland by 7 m. The advance was neither constant nor uniform. In dry times, there was very little change but in the wet months, the gullies advanced several feet due to rain water flowing into them. Throughout this period, the exposed slopes were gradually healing over. By the end of May 1962, the catchment area was well grassed and no erosion had taken place in the preceding wet season. When I arrived in the park some four years later, there were no signs of erosion and the gullies were stationary.

Paradoxically, the erosion gullies have improved the biodiversity of the area. Some of them are up to 15 m deep with quite a different microclimate at the bottom. Trees and bushes flourish within these sheltered trenches, providing improved habitats for birds and small antelopes as well as for leopards. The hippos themselves also benefit as the soil washed down from the gullies form alluvial fans extending into the water. Hippos love to haul out on these fans by day in order to indulge in some cautious sunbathing. The fans are also favoured by numerous other ungulates, attracted by the lush vegetation growing on them, as well as by water birds, which use them as loafing grounds for preening and sleeping. They proved to be important tourist attractions and were regularly visited by tourist launches.

An attempt to analyse more precisely the effects of grazing on plant growth was

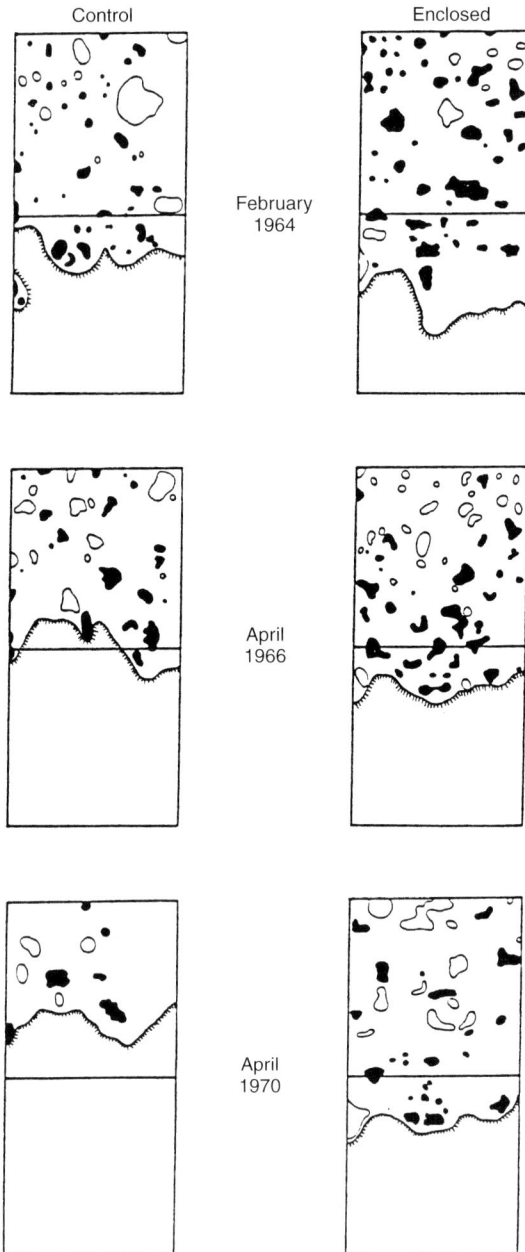

Figure 7.1 Plan of vegetation in plots inside (ungrazed) and outside (grazed) of grassland enclosures in a region of Queen Elizabeth National Park, Uganda, that was heavily grazed by hippos. The grazed plots act as a control. The movements of the steps (shallow erosion shelves) are much greater in the control plots than in those protected from grazing. Bases of plants of *Sporobolus stapfianus* are shown in black and other species are outlined. The steps are shown as hatched lines. From Lock (1972).

subsequently made by Lock (1972) in the Queen Elizabeth Park between 1964 and 1970. The technique used was to compare the grasses inside and outside ditched enclosures, which were placed in three vegetation zones at various distances of up to 3 miles (4.8km) from permanent water, well within the estimated grazing range of most of the hippos in the region. At the time, hippos comprised over 50% of the large mammal biomass. The next species in line, the elephant, was thought to be of limited significance as far as grassland changes were concerned and although other grazers were present, including buffalo, warthog, waterbuck and kob, most of the vegetational changes recorded were considered to be due primarily to hippos.

The enclosures were constructed at a time when heavy grazing by hippos was taking place so that subsequent changes within the enclosures represented the effects of the removal of grazing pressure. A clear change was a cessation in the spread of bare ground. One of the effects of heavy grazing is the production of shallow erosion shelves, known as steps, several centimetres in depth, which separate short grasslands from bare ground. Fig. 7.1 shows how these steps in unprotected plots move slowly over the short grassland, destroying it, whereas the steps are stationary within the ditched enclosures. Bare patches suffer from a lack of protection from heavy rain, which causes further erosion, and from run-off, which denies the benefit of moisture to the soil.

There were also changes in the species composition of the grasses with a reduction within the enclosures of tussock-forming grasses, primarily *Sporobolus pyramidalis*, which is favoured by grazing, and their replacement with long-stemmed grasses which although still tussock-forming, have leaves growing from the stalks. As an example, Table 7.3 shows the changes that occurred in one enclosure, number 4, which was sited in *Themeda triandra* grassland on black clay soil.

The depressing effects that hippos have on grassland are not confined to Uganda. O'Connor & Campbell (1986) measured the biomass, percentage cover and height of grasses on exclusion plots in the Gonorezhou National Park, Zimbabwe, and com-

Table 7.3 Changes in the species composition of grassland, 22 months after enclosure, on black clay soil in the Queen Elizabeth National Park, Uganda. Weights are means of clippings from 10 plots, each 100×50 cm in area. P – level of significance, NS – difference not significant. (From Lock, 1972.)

	Weight (g)		
	Inside Enclosure (ungrazed)	Outside Enclosure (grazed)	Significance
Grass Species			
Hyparrhenia filipendula	96.9 (31.0%)	71.1 (40.7%)	NS
Themeda triandra	93.1 (29.8%)	34.5 (19.7%)	NS
Bothriochloa sp.	60.3 (19.3%)	16.8 (9.6%)	$P<0.05$
Chloris gayana	25.8 (8.3%)	13.9 (7.9%)	NS
Sporobolus pyramidalis	7.2 (2.3%)	24.4 (14.0%)	$P<0.01$
Total dry matter	313	175	$P<0.001$
Total living matter	125	93	$P<0.01$
Total dead matter	188	82	$P<0.001$

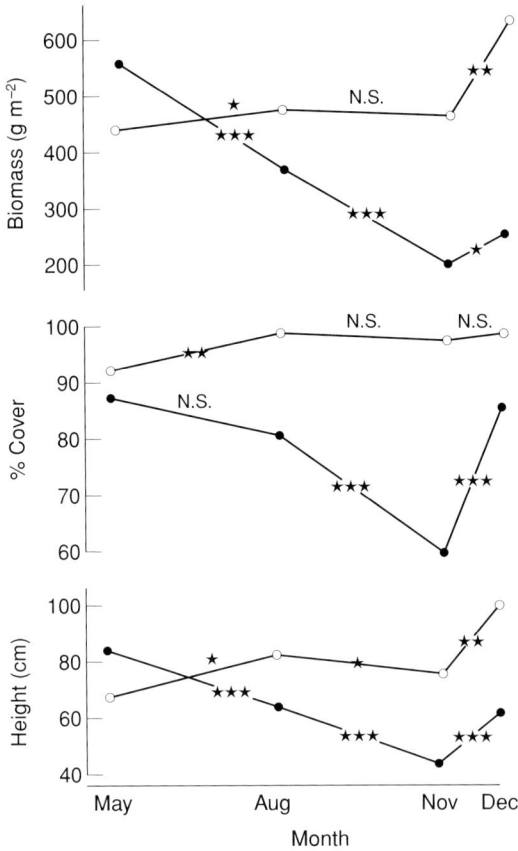

Figure 7.2 Changes in the vegetation following enclosure of plots in dune grassland in Gonorezhou National Park, Zimbabwe. Open circles – inside enclosure (ungrazed); closed circles – outside enclosure and subject to heavy hippo grazing. Significance of seasonal changes are shown by ★ $P<0.05$; ★★ $P<0.01$; ★★★ $P<0.001$; NS not significant. Note the improvement in the vegetation after only seven months of protection from grazing. From O'Connor & Campbell (1986).

pared them with measurements made on transects exposed to hippo grazing. After only seven months, significant differences were apparent as can be seen from Fig. 7.2, which shows the results for the dune grassland, the vegetation type most favoured by hippos. Similar patterns were found in comparisons made in other riverine vegetation.

The effects of hippo removal on other large mammals

It is to be expected that the consequences of such a large perturbation as the removal of as much as a third of the hippo population would have an effect on the other animals as well as on the vegetation. Indeed such changes must inevitably follow as it is the vegetation that determines the composition of the animal communities dependent upon it. Following on from Thornton's study of the plants on Mweya Peninsula, I undertook to look at the effects on the other large mammals. There would also have been consequences to other animals, such as birds and invertebrates, but unfortunately, there was nobody available to study them.

The numbers of large mammals on Mweya Peninsula in the past were well known from game counts carried out in 1956–57 by Petrides & Swank (1965) and between 1963–67 by Field & Laws (1970). My counts extended from 1968 to 1973 (Eltringham, 1974, 1980). After my departure from Uganda, the counts were continued until January 1976 by Yoaciel (1981). These counts, therefore, spanned the period from before the hippo cull to well after culling had ceased. The results are shown in Table 7.4.

The 1956–57 figures represent the community structure at a time when the Park was assumed to have an excessive number of hippos. All the hippos were cleared from the Peninsula study area in 1958, when 270 animals were culled.

Any that ventured onto the Peninsula after that date were shot. The culling ceased in 1966 so that for eight years the region was effectively free of hippos. (The average of one for the period 1963/67 does not mean that one was present throughout but it does show that some were trying to return even though the consequence was death.) Once the pressure was off, however, the hippos quickly returned with an average of over five to a square kilometre present within a year of culling being stopped. This illustrates a well known phenomenon termed the vacuum effect. In this case, the culling created a hippo vacuum, which was quickly filled by hippos moving in from outside the area. Within five years the population had recovered to above the pre-cull level. This rather nullifies the effects of culling and suggests that if hippo numbers are to be kept within agreed bounds, a one-off cull will not achieve the purpose and continuous shooting will be necessary.

It is interesting to note that although the hippo density on the Peninsula soon

Table 7.4 The number of large mammals (km^{-2}) on Mweya Peninsula between 1956 and 1975 before and after removal of hippopotamus in 1958. Numbers in italics represent percentages of the total number present. (After Petrides & Swank, 1965; Field & Laws, 1970; Eltringham, 1974, 1980 and Yoaciel, 1981.)

Date	1956/57	1963/67	1968	1969	1970	1971	1972	1973	1974	1975
Buffalo	4.6	27.8	40.4	30.7	20.3	24.5	26.8	22.8	15.6	11.0
	10.0	*49.9*	*57.2*	*54.5*	*39.8*	*39.6*	*40.5*	*33.2*	*28.7*	*24.0*
Bushbuck	3.7	3.1	2.4	1.7	2.1	2.9	2.4	2.9	2.0	1.8
	8.1	*5.6*	*3.4*	*3.0*	*4.1*	*4.7*	*3.6*	*4.2*	*3.7*	*3.9*
Elephant	2.2	3.5	5.8	2.5	3.6	2.4	2.1	1.9	1.8	2.3
	4.8	*6.3*	*8.2*	*4.4*	*7.1*	*3.9*	*3.2*	*2.8*	*3.3*	*5.0*
Hippo	21.3	1.0	5.4	7.2	11.0	17.5	22.2	25.3	27.5	26.6
	46.5	*1.8*	*7.7*	*12.8*	*21.6*	*28.3*	*33.5*	*36.8*	*50.7*	*58.1*
Warthog	9.4	8.6	6.7	5.7	5.1	6.1	5.1	7.3	5.6	3.0
	20.5	*15.4*	*9.5*	*10.1*	*10.0*	*9.9*	*7.7*	*10.6*	*10.3*	*6.6*
Waterbuck	4.6	11.7	9.9	8.5	8.9	8.4	7.6	8.5	1.8	1.1
	10.0	*21.0*	*14.0*	*15.1*	*17.5*	*13.6*	*11.5*	*12.4*	*3.3*	*2.4*
Total No.	**45.8**	**55.7**	**70.6**	**56.3**	**51.0**	**61.8**	**66.2**	**68.7**	**54.3**	**45.8**

recovered and indeed became higher than it was when severe habitat damage was being caused, there was little evidence of a similar problem with overgrazing. A possible contributory factor was the improved rainfall that occurred in the early 1960s, which may have ameliorated the effects of the herbivore pressure. Alternatively, the animal/plant balance may have moved to a new, but stable, equilibrium. The increase in hippo numbers followed a sigmoid curve, suggesting that the asymptote had been reached and that the population was settling down to a new carrying capacity not very different from what it was before the cull. The experiment could not be continued further since the 1970s witnessed a breakdown of law and order in Uganda, and the hippos in the Queen Elizabeth Park were indiscriminately slaughtered until only some two or three thousand of the original 14 000 remained.

These counts reveal some interesting consequences of the reduction of hippos on the other large mammal species. An inspection of Table 7.4 suggests that the effect has been greatest on buffalo numbers. This effect becomes more apparent if the counts for hippos and buffaloes are graphed, as in Fig. 7.3. The removal of hippos resulted in a massive increase in buffalo numbers, which continued to rise for a year or two after the culling was stopped. Once hippo numbers entered on their steep rise, however, the buffalo population went into a sharp decline, which, apart from a slight blip in the early 1970s, continued until the end of the observations. From this it seems that hippos and buffaloes experience an inverse relationship with each other.

There is no evidence of physical competition between the two species, which tend to meet only at wallows, but it may be that competitive exclusion operates through food (see p. 82) It is known from Field's work (Field, 1968c) that the feeding habits of the two are complementary, with the buffalo keeping back the tussock grasses to allow the creeping grasses, favoured by hippos, to spread. The hippos, in return, promote the growth of the buffaloes' favoured food plant, *Sporobolus*. It is possible that such facilitation operates only at low hippo densities and that at higher levels, competition takes over. The buffalo is one of the three species mentioned by Field (1972) as being possible competitors of the hippo. Neither of the other two potentially competing species, warthog and waterbuck, displayed any significant changes when the hippo numbers were showing their maximum rise.

Eltringham (1980) showed that, in addition to numbers, both the biomass and the energy consumption of the large mammal community of Mweya Peninsula increased as a result of the culling. The situation at the end of 1973 was more balanced than before or since, with neither hippo nor buffalo dominating the community, but by 1975 hippos were once again taking over. Any attempt to impose a balance through management, therefore, would involve an open-ended policy of hippo culling.

Conclusions from the hippo culling in Uganda

The hippo culling was a contentious issue in the 1950s and there are those who felt that it was unnecessary and that nature would take its course to restore the ecosystem to a balanced state. The fact that the vegetation recovered from its poor condition would seem to vindicate the cull but, as was pointed out, the improved rainfall may have been the more important factor in the recovery. One has to examine the dilemma faced by the managers in the 1950s who, without the benefit of hindsight,

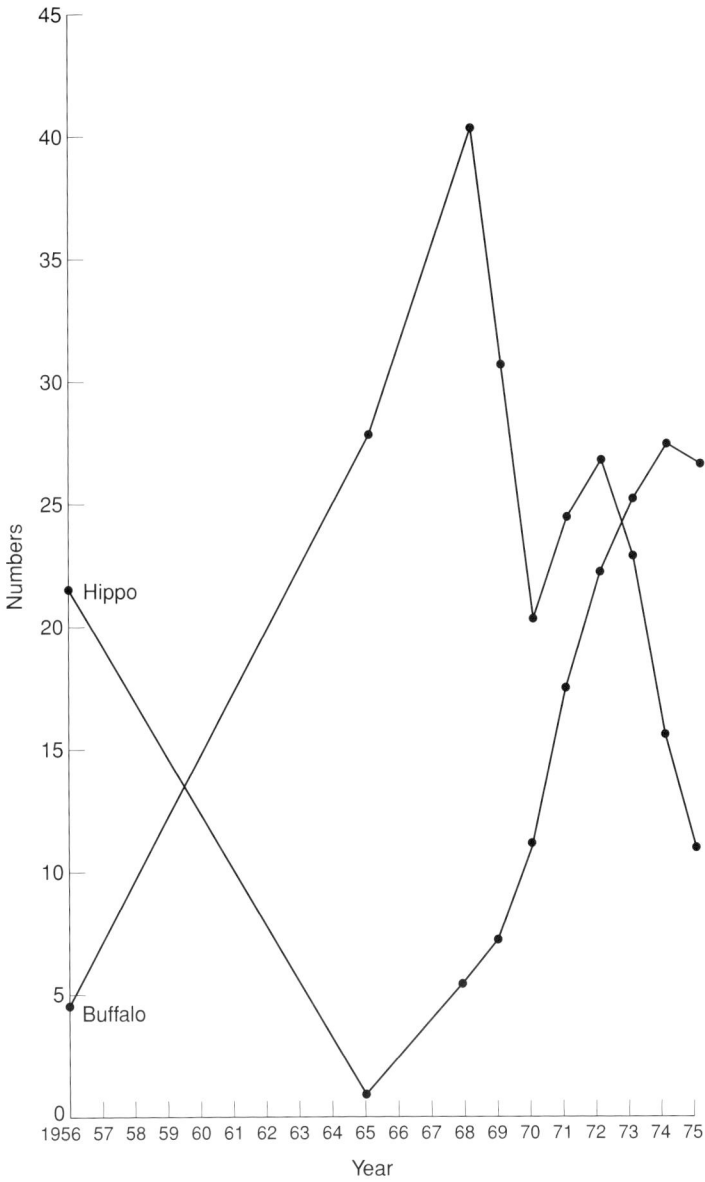

Figure 7.3 Inverse relationship between the numbers of buffalo and hippos on Mweya Peninsula, Queen Elizabeth National Park, Uganda, following the elimination of hippos in 1958. Hippos were kept off the Peninsula until 1966, after which the population was allowed to recover. See Table 7.4 for sources of information.

had to make the difficult decision. The worst-case scenario was the reduction of the park to a semi-desert, with the loss of many animal and plant species and the managers very properly thought that this was too great a risk to take. By removing, as they thought, half the population, there would be an immediate relaxation of grazing pressure, and even if it proved to be ineffective, there would still be a large number of hippos remaining. In other words, no long-term harm would have been caused. We shall never know if the dire forecast would have materialised, and nature may well have taken its course but in a way that would probably have offended the sensibilities of those who objected to killing animals. Under more natural conditions, the hippos might have moved away but in the park there was nowhere to move to and it is more likely that they would have starved.

What this exercise has shown, however, is that a single cull will not solve the problem. This is accepted nowadays on theoretical grounds and in future management interference at this level would have to be continued indefinitely in order to be effective. Although it is difficult to measure, there is such a thing as a carrying capacity and a population will return to it after a perturbation unless draconian measures are taken to stop it doing so. It may well be better to let matters take their course and accept whatever ecosystem develops. In the event, financial considerations usually prevail. The other principal lesson to be learned is that long-term planning will not necessarily come to fruition particularly in unstable countries such as Uganda then was. If the managers of the 1950s had known that less than 20 years later the country would collapse into anarchy with wholesale slaughter of hippos, they may well have reached other conclusions.

POPULATION ECOLOGY OF HIPPOS

Population structure

The structure of a population refers to the relative proportions of each age and sex class present. It is useful in that it indicates the "health" of the population, e.g. a low number in the younger age class would suggest poor recruitment or a high juvenile mortality rate. The typical age struture of a large mammal population (Fig. 7.4) is one with many individuals in the infant age class with a sharp fall in numbers of juveniles and subadults. Once the high juvenile mortality has had its effect, there is only a slow decline in numbers in the succeeding age classes until extreme old age, when there is a sharp decline to zero when all animals are dead. Separate structures for each sex need to be drawn up for most species as the female usually lives for a much longer time than the male.

The population structure allows one to draw up a life table similar to those constructed by actuaries to decide life insurance rates. They can in theory also be used to estimate the maximum sustainable yield that can be taken from a population but the data have to be very accurate for this to be of any practicable use. Life tables may be based either on the age of death of individuals (time-specific tables) or on the age composition of the population at a particular point in time (age-specific tables). In practice, of course, observations are made not on the entire population but on a representative sample. Data for time-specific tables of large wild mammals are usually

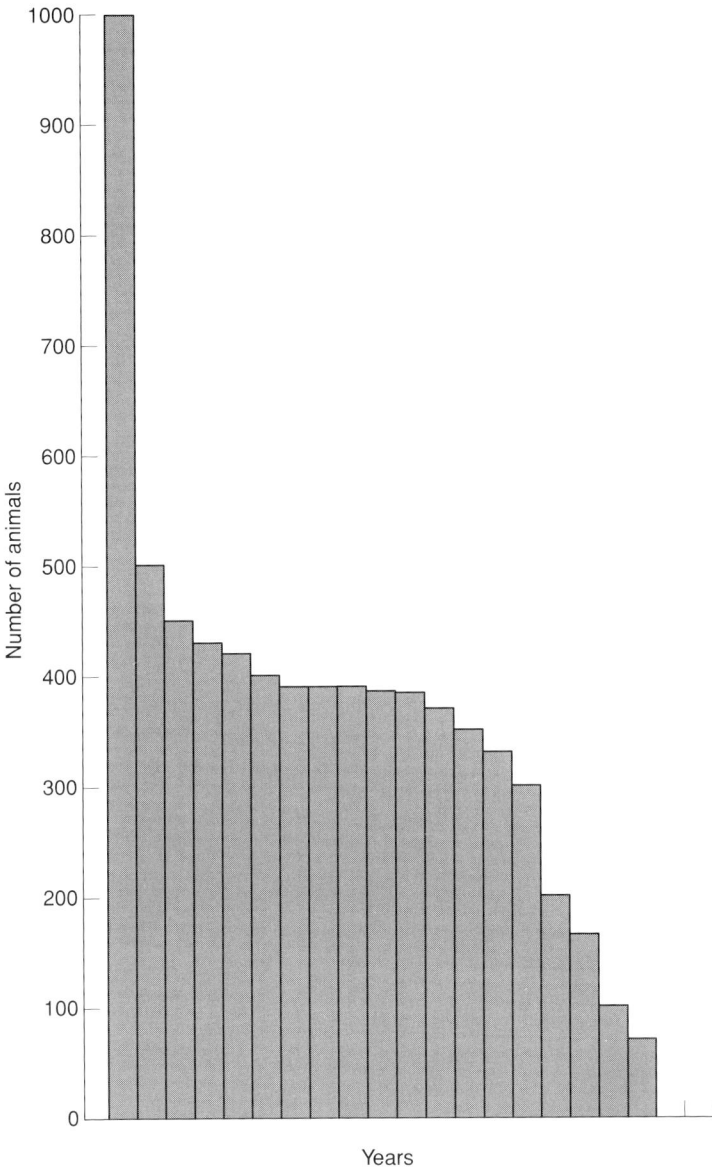

Figure 7.4 Population structure of a typical large mammal. There is an initial high juvenile mortality followed by a middle age of low mortality, which does not increase much until old age is reached.

obtained from skulls or other remains collected in the bush, whose age at death can be estimated from tooth wear etc. Age-specific tables can be constructed simply by observing how many animals there are in each age group. This is difficult, if not impossible, with living animals as their ages cannot usually be determined in the wild and such life tables have to be based on a sample of shot animals, whose ages can be estimated with some degree of accuracy.

Both methods have their limitations. Skulls of young animals are less durable than those of mature adults and therefore the younger age classes are usually under-represented in samples used for time-specific life tables. Young animals are also less likely to be included in samples taken for age-specific tables. As a consequence, arbitrary values are often given to these classes.

Laws (1968a) has calculated time-specific life tables from lower jaws collected in the Queen Elizabeth National Park in Uganda. Depending on the assumptions made, two tables were drawn up showing life expectancy at birth to be 10.1 years and 9.3 years respectively. Maximum survival rose to 18.7 years at 2 years of age according to one model and to 24 years at 3 years of age in the other. Thereafter there was a steady decline throughout life. The survivorship curves constructed by Laws from these analyses do not depart significantly from that of a generalised large mammal. A much simpler analysis of lower jaws found in the adjoining Albert National Park in Zaire by Bourlière & Verschuren (1960) revealed a similar survival curve although it did not reveal the infant mortality found in the Ugandan data. This is probably due to the low number of calves in the sample.

Population structures of hippos based on age-specific data have been provided by a number of authors. One such is that by Marshall & Sayer (1976) for hippos culled on the Luangwa River, Zambia, between 1965 and 1971. Their figures (reproduced here as Fig. 7.5) lump the sexes together but their tables for the 1970 and 1971 culls separate the sexes enabling the data to be plotted for each sex. This has been done in Figs 7.6 and 7.7 for males and females respectively. The samples are clearly biased as young animals were much less likely to be shot and are under-represented.

These distributions are far from typical of the standard pattern shown in Fig. 7.4, even allowing for the poor representation of younger age classes. Some of the difference may be due to inaccurate age estimates and some to the unrepresentative nature of the culled sample but others may have been caused by differential mortality acting throughout the lives of these long-lived creatures. Theoretically, there should always be fewer individuals in an older age group but this is frequently not the case, suggesting that at some time in the past, there was heavy mortality that affected only a particular age group or groups. No definite conclusions can be drawn from this analysis although Marshall & Sayer draw attention to the generally low number of individuals in age classes XI–XIV (20–27 years) and the large number in age class XV (30 years). They attribute the gap to inaccuracies in age estimation due to the highly seasonal climate in Zambia relative to that in Uganda, where the technique was developed by Laws (1968a). The drier conditions would lead to a greater amount of soil being eaten along with the grass so increasing the wear on teeth. This would result in specimens being allocated to older age groups than those to which they really belonged.

Suzuki & Imae (1996) provide the age and sex composition of 457 hippos culled in the Luangwa River, Zambia, a quarter of a century after the survey there by Marshall & Sayer. Their results (Figs 7.8 and 7.9) show a similar dearth of young adults although there were plenty of females in the mid to late twenties. The curve for males is not unlike that for the typical large mammal apart from the low numbers in the young stages.

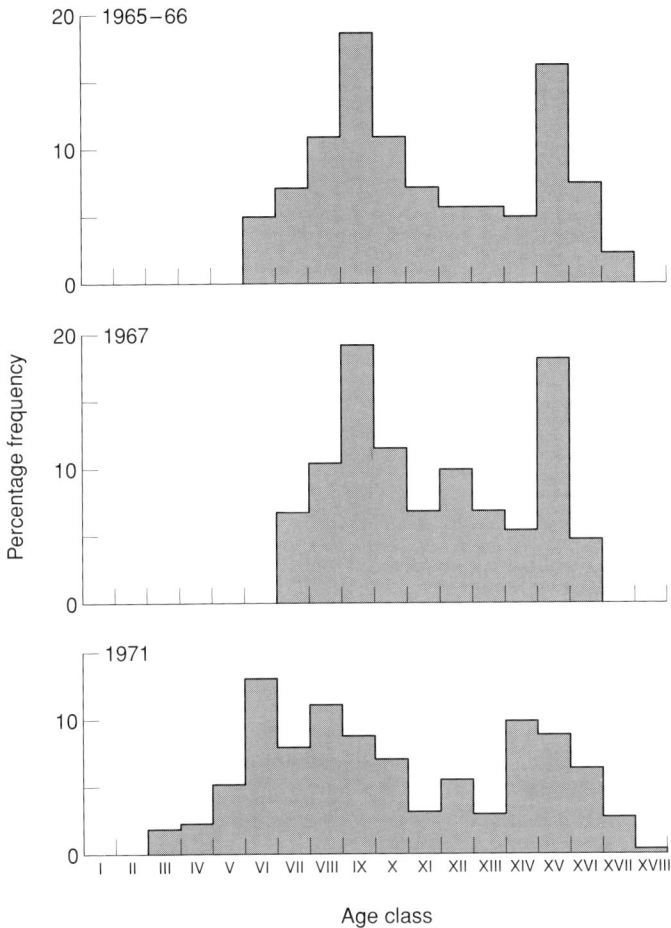

Figure 7.5 Population structure of common hippopotamus culled on the Luangwa River, Zambia, between 1965 and 1971. Mixed sexes. From Marshall & Sayer (1976).

Fluctuations in population size

Very few attempts have been made to follow the size of a population of hippos over a long period. This is perhaps not surprising given the longevity of the species and the short working life of researchers. Local counts have often been made over short stretches of rivers etc. but little of biological interest has emerged. Changes in numbers can either be readily explained or shown to be due to inconsistencies in the counting technique. Marshall & Sayer (1976) plotted the numbers of hippos over a ten-year period in their cropping area in Zambia and showed that fluctuations in population size tended to follow the numbers removed by cropping. There was nevertheless a large, inexplicable increase between 1968 and 1969 which could have been

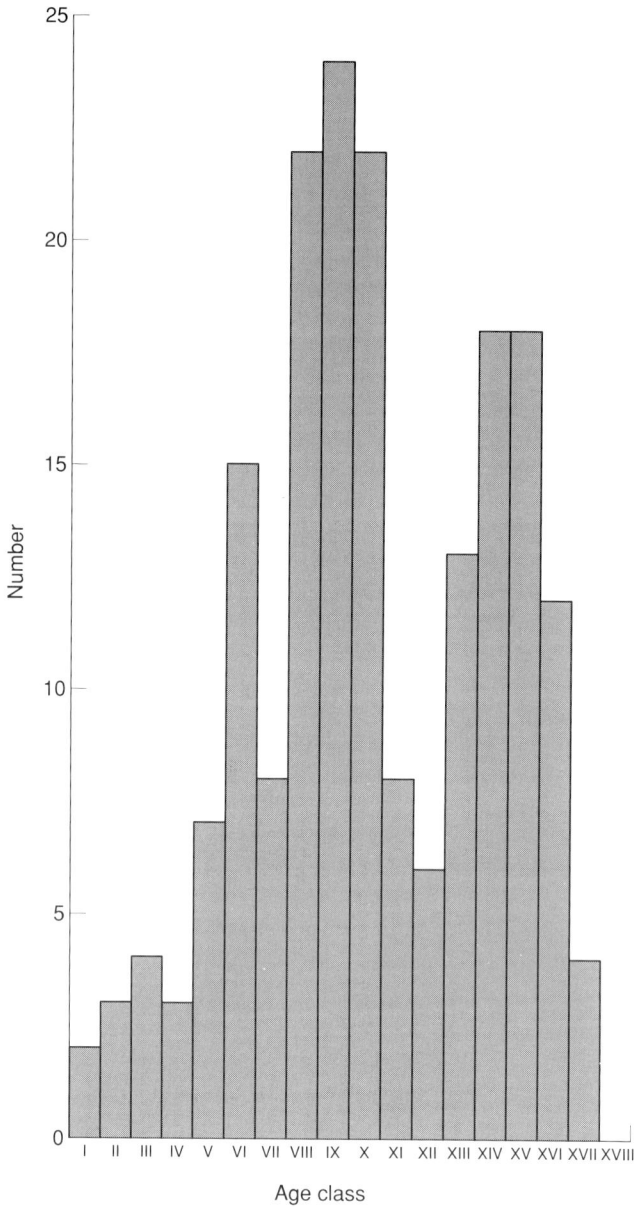

Figure 7.6 Population structure of 189 male common hippopotamus culled in Zambia in 1970. After Marshall & Sayer (1976).

due to inaccurate counting or to an influx of hippos into the relatively short stretch of river that was surveyed. Viljoen & Biggs (1998) found some variations in numbers on three rivers in the Kruger National Park over an 11-year period but overall there was no trend.

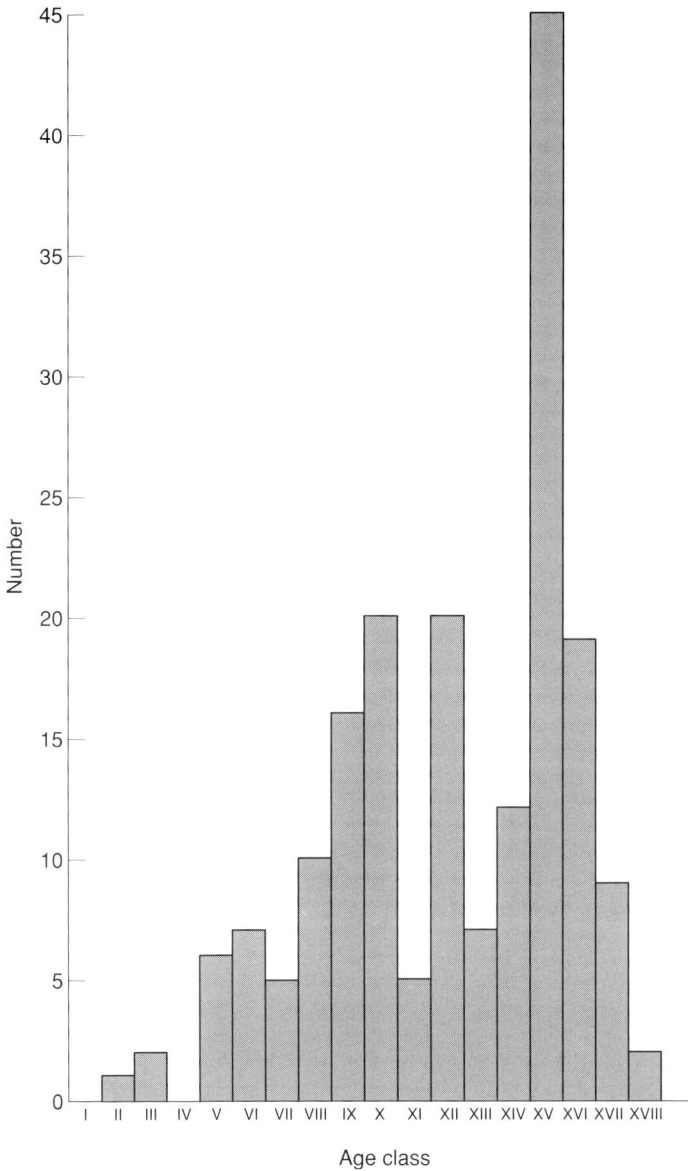

Figure 7.7 Population structure of 186 female common hippopotamus culled in Zambia in 1970. After Marshall & Sayer (1976).

Some counts made over a longer period and covering a wider area are those made by the Nuffield Unit of Tropical Animal Ecology and its successor organisation, the Uganda Institute of Ecology. These data have not been published but are available in the annual reports of the research station. The technique used was aerial survey with

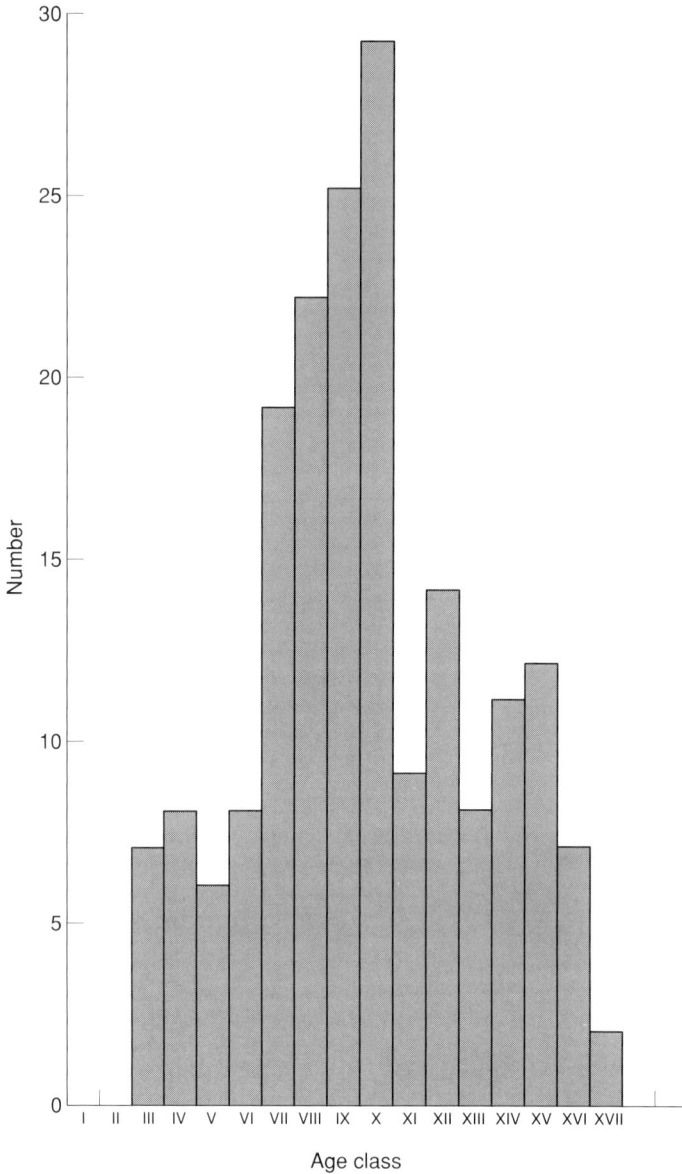

Figure 7.8 Population structure of 187 male common hippopotamus culled in Zambia in 1995. After Suzuki & Imae (1996).

photography of the larger aggregations, a method that tends to under-estimate numbers. The counts covered the whole of the Queen Elizabeth National Park in Uganda including that part of the shore line of Lake George which is not in the park. The results are shown in Fig. 7.10. The initial population of 12 393 in 1960 had declined

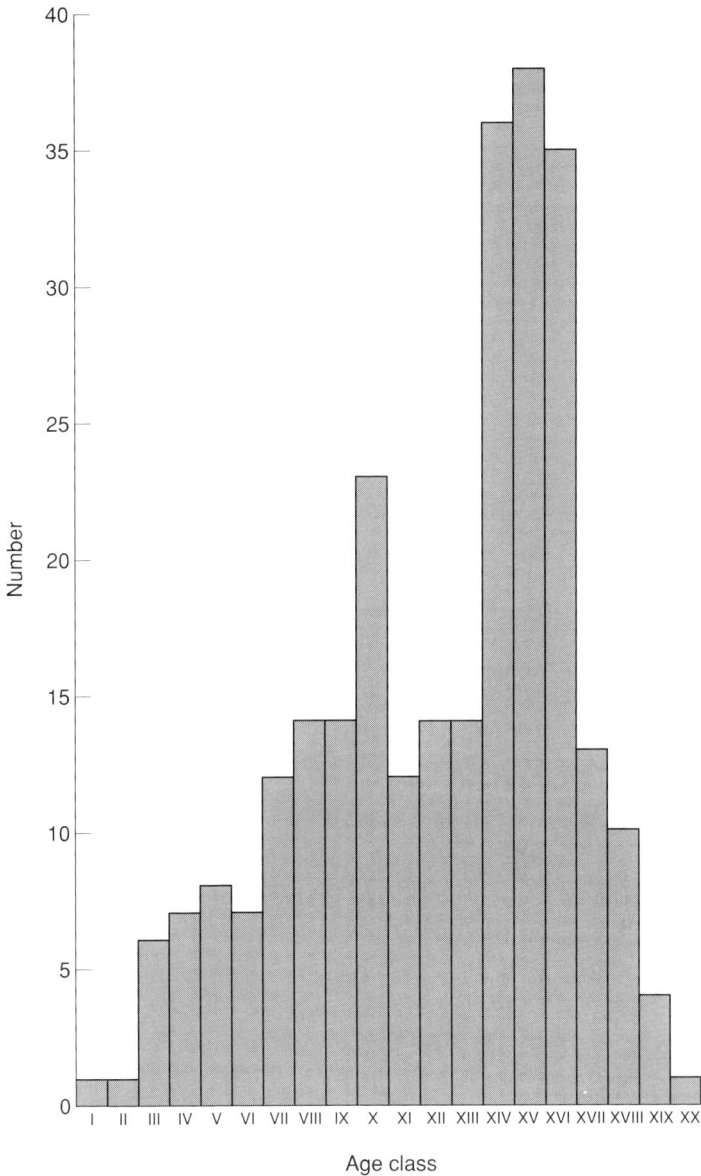

Figure 7.9 Population structure of 270 female common hippopotamus culled in Zambia in 1995. After Suzuki & Imae (1996).

to 8005 by May 1964, a drop of 4388. This is close to the number culled during this period, which was 4485. Culling ceased soon afterwards and numbers climbed back to around the original total. Subsequent counts showed a slight decline although this could be an artefact due to inaccuracies inherent in the technique. This tendency for

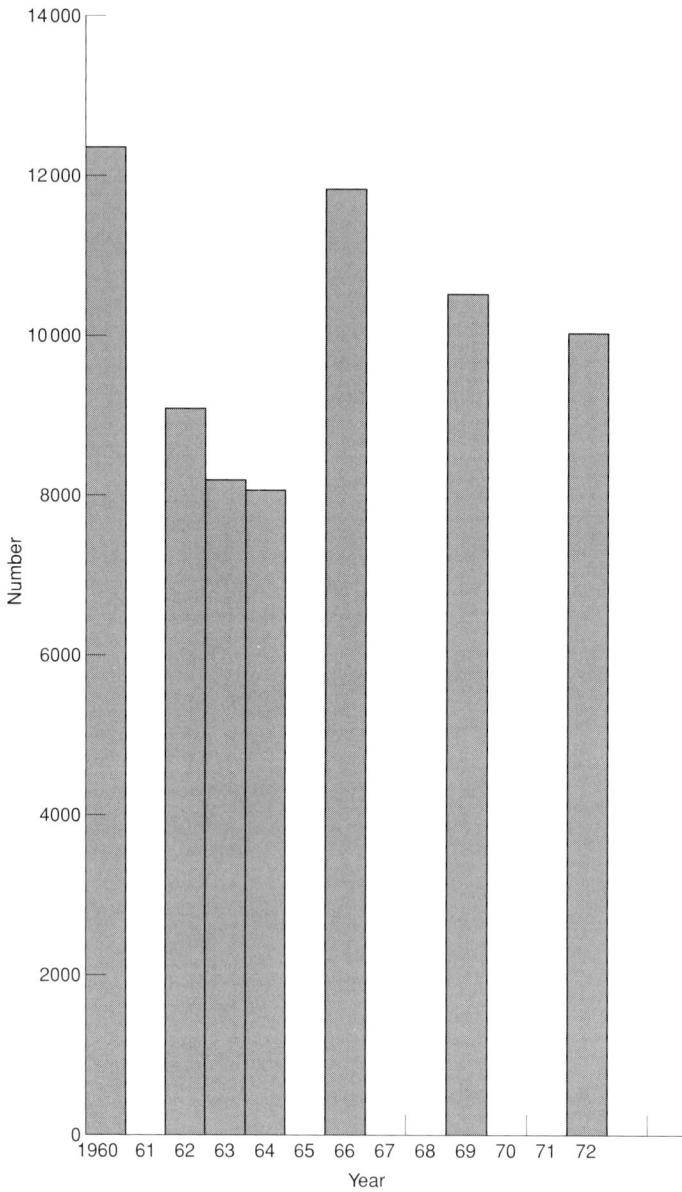

Figure 7.10 Numbers of common hippopotamus counted from the air in Queen Elizabeth National Park, Uganda between 1960 and 1972. The counts include that part of the shore line of Lake George that is not within the park. Data from the annual reports of the Nuffield Unit of Tropical Animal Ecology.

hippo numbers to return to their previous level on the cessation of culling is a further illustration of the vacuum effect mentioned earlier in this chapter. The population plummeted to around 3000 or less a few years later but there was no mystery about the cause as by then, President Amin's malevolent rule had begun to take effect and hippos were slaughtered wholesale for their meat. An aerial count made in 1989 returned a total of only 2172 (R. Olivier, *in litt.*). It remains to be seen if numbers will climb back up to their previous levels.

CHAPTER 8

DISEASES, PARASITES AND COMMENSALS OF HIPPOS

*Armed truce between crocodiles
and hippos*

Hippos are as vulnerable to diseases and parasites as any other wild animal but they have also evolved relationships with other species that are neither pathogenic nor parasitic. Such partners are called commensals, literally feeding at the same table. Commensals do no harm to each other but do no good either. If there is a mutually beneficial consequence from the partnership, the relationship becomes a symbiotic one. It is not always easy to distinguish between commensalism and symbiosis if not much is known about the biology of the species concerned for there may be benefits that are not apparent to the human observer. Parasites cannot always be distinguished from commensals either for commensals can become parasitic if, for example, one of the partners loses condition. The distinction between parasites and disease organisms is often a semantic one but there seems to be a general understanding that viruses, bacteria and protozoans are disease organisms and anything bigger is a parasite. Parasites may also be vectors of diseases as in the case of mosquitoes, which spread the proto-zoan *Plasmodium*, the causal agent of malaria.

110

DISEASES OF HIPPOS

There have not been many studies of disease in hippos as it is not a topic that interests many people, apart from zoo keepers. Hippos do not make ideal experimental animals and a pathologist is likely to look for something more mouse-sized for a study of the effectiveness of vaccines etc. Nevertheless, there have been some zoo studies as well as studies of wild animals, particularly those shot in culling operations or caught through drug immobilisation.

Viral diseases

One of the first in-depth studies of disease in hippos came from the examination of specimens taken during the hippo culling in Uganda in the 1950s and '60s (p. 89). Blood samples were drawn from the carcases soon after death and tested for antibodies (Plowright *et al.*, 1964). One of the diseases investigated was rinderpest. This European cattle disease was introduced into Africa in the 19th century and devastated the wildlife as well as the native cattle. Not all species are susceptible to rinderpest and the hippo was assumed to be immune because no case had been reported. It is more likely that the disease had never been looked for despite its known existence in pigs and peccaries, which are assumed to be close relatives of the hippo. When the animals culled in Uganda were examined, however, antibodies to rinderpest were found to be

Table 8.1 The occurrence of rinderpest-neutralising antibody in the sera of 315 Ugandan hippos shot in 1962. (From Plowright *et al.*, 1964).

Age Class	Estimated age (years)	No. tested	No. & % positive	Lumped Data Tested	Positive (%)
3	1	3	0 (0%)		
4	3	9	0 (0%)		
5	4	31	0 (0%)		
6	7	7	0 (0%)	70	0 (0%)
7	8	21	0 (0%)		
8	11	29	0 (0%)		
9	13	35	1 (3%)		
10	15	26	1 (4%)		
11	18	13	1 (8%)		
12	20	9	0 (0%)	120	5 (4%)
13	22	18	2 (11%)		
14	25	19	0 (0%)		
15	28	50	13 (26%)		
16	31	42	16 (38%)		
17	33	16	8 (50%)		
18	30	1	0 (0%)	115	41 (36%)
19	38	4	3 (75%)		
20	41	2	1 (50%)		
TOTALS		**315**	**46 (15%)**		

widely present although none of the animals examined showed obvious clinical symptoms of the disease.

Sera drawn from the blood of 315 hippos shot in 1962 were tested and 46 were found to be positive to rinderpest-neutralising antibodies (Table 8.1). There were marked differences between the age classes in the possession of antibodies, reflecting the occurrence of rinderpest outbreaks in the past in the region where the hippos were shot. The timing of these outbreaks are known from reports of the disease in other animals, from which it can be concluded that the hippos were exposed to rinderpest virus in 1920/21, 1931/33 and 1944/45. Hence hippos that would have encountered the virus for the first time would, at the time of sampling (1962) have been either 41–42, 31–33 or 17–18 years old. These age ranges correspond roughly to the ages of the animals showing a positive reaction (Table 8.1) all of which would have been alive at the time of one or other of the epizootics with the exception of the 9 to 15-year-old hippos, which were too young to have been exposed to the virus. Plowright *et al.* (1964) have no explanation for the presence of antibodies in these young animals other than that there might have been mislabelling of the samples or errors in estimating age. There is, of course, the possibility that there was a later, minor outbreak of the disease that was overlooked.

A surprising aspect of this study is the length of time that the antibodies persisted in the body and remained at high levels. With many viruses the mean titre of antibody falls during life although in some cases, such as measles in man, there is only a small decline over periods as long as 30 years.

Incidentally, this study provides a nice measure of the longevity of hippos in the wild. Assuming that the oldest animal in the sample (age group 20) was infected in 1920/21, it would have been 41 or 42 years old by 1962. This particular animal had teeth that were almost completely worn down and it was probably approaching the end of its life span.

It should be stressed that there is no evidence that the hippos actually suffered from rinderpest. It was mentioned earlier that no animal in the sample was showing clinical signs of the disease and there are no records in the literature of the disease affecting hippos. There have been reports of multiple deaths in hippos but suspicion has usually fallen on anthrax as the cause of death.

Bacterial diseases

Anthrax is a particularly virulent disease which affects most, if not all, mammals, including man. It is caused by a bacterium, *Bacillus anthracis*. There have been reports of anthrax in hippos and it is probably the most serious disease as far as an individual hippo is concerned although the evidence for its presence seems inconclusive. It is not easy to confirm the occurrence of anthrax, unless a blood sample from the victim can be examined very soon after death, for the bacteria of decomposition are very similar in appearance to anthrax bacilli and specialised tests are required, which are usually impracticable in the bush. Clinical signs, on the other hand, are often fairly obvious, with tarry blood oozing from all orifices.

Anthrax bacilli are capable of surviving for years in the soil as cysts, and can infect grazing animals by entering through minor scratches in the lips caused by soil particles, particularly in times of drought, when the herbivores have to graze close to the ground. Suitable conditions for the germination of the spores include warm ambient temperatures of around 27°C and soils high in organic matter with an alkalinity of about pH 7.5. These ideal conditions are realised in the vicinity of waterholes and salt licks, which are frequented by many large mammal species. Hence the disease can be spread very easily among the large mammal community, especially in the dry season. The hippo's life-style lends itself to the transmission of disease organisms, involving, as it does, close bodily contact in often muddy, faecal-rich water. However, anthrax is unlikely to be an important mortality factor in hippo population dynamics because of its extreme morbidity, which quickly leads to the death of infected animals before they can pass on the disease to many others.

A severe outbreak of anthrax in the Luangwa Valley, Zambia, has been documented by Turnbull *et al.* (1991). An abnormal mortality in the hippos was noticed in 1987, when over 1420 deaths were tallied along a 167 km stretch of the Luangwa River from June until November, when the disease died out with the onset of the rains. By extrapolation, the total death toll in the whole population was estimated to be in excess of 4000. Anthrax was established as the cause of the deaths. Many other species were affected, particularly buffalo and elephant, but giraffe, puku, kudu, waterbuck, wild dog and baboon also suffered. The average death rate in the hippos was 21%, ranging from 5.7 to 55.5% in the various schools. A smaller outbreak occurred about 100 km upstream in June of the following year, reaching a peak in August and September.

Guilbride *et al.* (1962) looked for the presence of anthrax in hippos culled in the wallows area of the Queen Elizabeth National Park, Uganda. Blood smears were taken from 287 animals but all proved to be negative. The intestinal contents of 65 hippos were also examined for anthrax but again no pathogens were found. There was no trace of spores in 46 mud samples taken from the wallows and attempts to isolate *B. anthracis* from four bones collected from the lake shore, where an outbreak of anthrax had occurred in hippos two years previously, proved unsuccessful. These observations bear out the comment made above that anthrax is so severe that it is self-limiting.

Another bacterial disease organism found in hippos is *Salmonella*, which is responsible for many outbreaks of food poisoning in the human population. Guilbride *et al.* (1962) examined the intestinal contents of 149 culled hippos and found *Salmonella*, belonging to five different species, in ten of them. The only hippo that was visibly sick was carrying *S. typhimurium*. The bacterium is very common in cattle and other animals in this part of Uganda so it is not surprising to find it in hippos. Although *Salmonella* can be very debilitating, its presence does not necessarily mean that the host is actually suffering from disease.

Brucellosis is a highly contagious disease caused by the bacterium *Brucella abortus*. As its name suggests, it is the causal agent in contagious abortion in cattle. It is spread in a variety of ways but it is usually transmitted through urine contamination. It can also be spread by biting insects and ticks as well as by the common house fly. Like many other wildlife diseases, it is a zoonosis, i.e. it can pass between man and animals. Guilbride *et al.* (1962) examined serum samples from 144 hippos and isolated *Brucella* from eight with a further 17 suspicious cases.

Protozoan diseases

Trypanosomes are flagellate protozoa parasitic in the blood stream. They cause Chagas' disease and sleeping sickness in man and a series of diseases in animals collectively known as trypanosomiasis. They are usually spread in Africa through blood-sucking insects of the genus *Glossina*, commonly known as tsetse flies. Wildlife develops an immunity to the disease but domestic animals are severely affected, so much so that cattle cannot be kept in tsetse country. The fly has been credited with being the most effective conservationist in Africa in preventing the encroachment of cattle onto pristine wildlife habitat. Hippos suffer from the bites of G. *brevipalpis* (Baker, 1969) and can be infected by trypanosomes, for *Trypanosome congolense* was found in the blood of one hippo by Guilbride *et al.* (1962). Garnham (1960) found two infected hippos from the same population out of a total of 137 examined and suggested that the parasite was *T. simiae*, which belongs to the *congolense* group. He suggested that the hippo might be a reservoir for *T. simiae*. The occurrence of the actual disease of trypanosomiasis has not yet been demonstrated in hippos.

Malaria is the most common (and deadly) protozoan disease in people but it has not been reported in hippos. This does not mean that it and other similar diseases do not occur as few hippos have been examined for their presence.

PARASITES

As mentioned above, parasites are taken to be multicellular organisms whose presence in or on an animal is detrimental to the host. Almost all phyla include parasitic forms and some are exclusively parasitic. Probably all animals have parasites and the hippo is no exception.

Endoparasites

Endoparasites live inside the body and feed on its tissues. As with all parasites, the life cycle must contain a mechanism for getting from one animal to another to prevent the parasitic population from becoming extinct with the death of the host. In the case of gut parasites, a common method is to pass eggs to the outside world in the faeces of the host. Sometimes the parasites pass directly to another host when the eggs or larvae are inadvertently eaten by an animal of the same species but other parasites employ a second species, such as an insect, to convey them to a fresh host. A common example is the mosquito, which sucks up blood parasites when it bites its victim. Such an agent is known as a vector, which may or may not suffer itself from the presence of the parasite. Some parasite cycles have two, or even three, vectors.

Platyhelminthes

The Phylum Platyhelminthes, or flatworms, have free-living forms, such as the familiar planarian "worms", but most are parasitic. The latter include the trematodes, or flukes, and the cestodes, or tapeworms. Trematodes are usually endoparasites occurring in the blood vessels or gut of vertebrates. Those belonging to one order, the Monogenea, have only one host but members of the other order, Digenea, use two or more secondary hosts to pass from one animal to another. A well known example is the liver fluke *Fasciola hepatica*. Tapeworms generally parasitise the gut as adults, with the larval forms occurring as cysts in the muscles of the secondary host, which is usually the prey of the primary host. The parasite is passed from one to the other when the prey is eaten.

Guilbride *et al.* (1962) found that there were surprisingly few helminths in hippos compared with the infestations found in domestic animals but members of the Paramphistomatidae, probably *Nilocotyle* and *Buxifrons*, were common in the forestomach and in the first one and a half metres of the small intestine. Ova of these flukes were also common. The liver fluke *Fasciola nyansae* was found in the livers of most hippos. It was not present in the large numbers generally associated with cattle but the bile duct in hippos is narrower than that in cattle making it more difficult to detect the parasites. Flukes were more common in young animals than in older specimens suggesting that the hippo acquires an immunity to infection with age.

Blood flukes (*Schistosoma* spp.) inhabit the hepatic portal system and the pelvic veins of birds and mammals. They can cause debilitating illnesses in man. They are unusual among flukes in being sexually dimorphic and for remaining more or less permanently *in copula*. Although they are blood parasites, schistosomes are not spread by bloodsucking insects. Instead the egg, usually in a spiny capsule, is laid in a capillary from where it migrates to the intestine or bladder to be passed to the outside world in faeces or urine. The eggs hatch out into a miracidium larva which penetrates a water snail and develops into a cercaria larva. This is released into the water and penetrates the definitive host to complete the life cycle. Infection, therefore, is through larvae swimming in the water and it is to be expected that hippos are infected. Schistosomes do indeed occur in hippos (Pitchford & Visser, 1981) although the extent to which they incapacitate the host is not clear. Mention has already been made (p. 30) of the atrophied ovaries in a Kruger hippo, which were possibly caused by blood flukes. The schistosome found in hippos has been shown from its ribosomal RNA to be a distinct species, *S. hippopotami* (Despres *et al.*, 1995).

Cestodes were rare in the hippos sampled by Guilbride *et al.* One cestoid egg was seen, probably from *Moniezia*, as a damaged part of an adult tapeworm resembling that genus was found in one individual. This is a genus present in sheep and cattle as well as in other large mammals.

Nematodes

Nematodes are ubiquitous animals found in every conceivable habitat both as parasites and as free-living forms. They are worm-like in shape and usually small although some parasites grow to a large size. Many of the parasitic species are found in the gut lumen of vertebrates and are passed from one host to another as eggs in the faeces. On hatching, the larvae infect a fresh host, often by burrowing through the skin. They make their way back to the gut in what has been called "larval wanderings" during which they pass through various organs of the host causing considerable damage on the way. Other parasitic nematodes pass through a secondary host or are spread by blood sucking insects. Several of the parasitic forms, such as hookworms, cause serious illness in man, including anaemia. Elephantiasis is also a nematode disease and is caused by filaria larvae blocking the lymph system.

No parasitic adult nematodes were found in the 25 hippos examined by Guilbride *et al.* (1962) but strongyloid ova were frequently seen so it is likely that the hippo is attacked by these parasites. It would be unusual if they did not harbour a variety of nematodes as all other large mammals do.

Ectoparasites

Hippos are widely infected with a monogene fluke, *Oculotrema hippopotami*, that lives on the outer surface of the eye. It does not appear to cause serious harm to the eye but it does seem to produce local irritation judging from the increased frequency with which an affected hippo blinks. The adult flukes attach themselves to the inner edge of the nictitating membrane of living hippos but they can be found under the eyelid in dead animals (Thurston, 1968a) so they possibly occur there in living hippos as well. Of the 42 hippos examined by Thurston, all but four harboured eye flukes with a mean number of parasites per hippo of 8.3 and a maximum of 41. Sometimes only one eye was affected and of 28 hippos for which the affected eye was recorded, six (21%) were parasitised in the left eye only and six in the right only. Both adult and immature parasites may be found in the same eye suggesting that there is no strong immune response. Nevertheless there is evidence that the parasite load decreases with age of the host in that a significant inverse correlation was found in the data for one month between the age of hippos and the number of infesting flukes, although the significance disappears if a second month of observations is included in the analysis.

  Given their aquatic habits, it is hardly surprising to learn that hippos are often infested with leeches, particularly in the ano-genital region. Apart from the loss of blood, the hippo does not seem to suffer unduly from their attentions. Guilbride *et al.* (1962) mention *Limnatis nilotica* as being particularly numerous.

  Guilbride *et al.* also recorded two species of ticks on the 67 hippos they examined with 358 specimens, mainly males, of *Rhipicephalus simus simus* being collected. The other tick species mentioned was *Amblyomma tholloni* with 150 records including 105 nymphs.

METABOLIC DISEASES

Not many metabolic diseases have been identified in the hippo, probably because carcases have rarely been subjected to veterinary analysis. Mention has already been made (p. 29) of the testis cords in the ovary of a lactating female from Uganda and of a teratoma in the enlarged ovary of another hippo from South Africa. These were certainly anomalies if not perhaps diseases. Many kidney stones were reported by Suzuki (1997) in an elderly female (*c.* 35 years old) culled in Zambia. After the stones had been removed, the affected kidney was found to be about half the weight of the other (465 vs. 930 g).

TRAUMA

The main trauma suffered by hippos comes from fighting, mainly between males over territorial rights, and most mature males are covered with scars to a greater or lesser extent. As mentioned earlier, severe gashes can be inflicted on opponents during these fights and even the victor rarely escapes unhurt. Sometimes the wounds are serious enough to cause death but this rarely occurs during the fight itself and it may be several days before the wounded animal succumbs. On the whole, wounds heal up perfectly well but the presence of pus in the fresh scars of a mature male culled in Zambia was reported by Suzuki (1997). The tip of a hippo's tusk was found under the skin of this animal, leaving no doubt over the cause of the wounds. The only other main cause of trauma in the hippo is human aggression, which is considered in the next chapter.

RELATIONS WITH OTHER VERTEBRATES

This section considers commensalism between hippos and other vertebrates as well as antagonistic contacts. Man is excluded because a whole chapter is devoted to the relationship between these two species.

Predation

The common hippo is not at risk from predators to any great extent, although there have been reports of lions ganging up on a lone hippo and, by repeatedly scratching it, eventually causing its death through loss of blood. The lions would have to be considerably nimble to dodge the hippo's counter-attacks. In all the contacts I have ever witnessed between the two species, the hippos have shown complete indifference to the presence of lions, which usually take care to move out of their way. Nevertheless, Ruggiero (1991) describes an adult male being attacked by lions in the Manovo-Gounda-St Floris National Park, Central African Republic. The hippo was later found dead but it was not certain that the lions were responsible although they were eating it. Guggisberg (1961) describes an incident in which lions killed two adult hippos by knocking them off their feet and grasping them by the throat. Green (1997) reports a young hippo being killed by lions in Benin in December 1973 and an adult succumbing to five lions in 1978 although he suspected that the hippo was ill.

Crocodiles are not predators of hippos although they will readily scavenge a dead one. Kofron (1993) observed the interactions between hippos and crocodiles in the Chipinda Pools of the Runde River in the Gonarezhou National Park in Zimbabwe. In all interactions the hippos were dominant. He recognised two types of confrontation, one in the water and the other on land. Such disputes usually occurred during the dry season when the river is reduced to a series of pools, which then have to be shared by both species. Hippos in the water do not tolerate the presence of crocodiles within 2 m and crocodiles avoid hippos by diving and swimming under them. The hippos threaten the crocs first by snorts, then by charges and if these fail, contact is made and the crocodile is shoved out of the way.

Confrontation on land is usually over the possession of basking sites during the cool season in July. These are in short supply and there is competition for them. Crocodiles bask as a group and on at least eight occasions, Kofron witnessed hippos displacing crocs from the basking grounds. A hippo does so by approaching slowly from the rear with the nose held near the ground until it is touching the crocodile's tail. The crocodile responds by lifting its head to an alert position and gaping or by moving its tail.

Eventually the hippo pushes the croc out of the way and settles in its place. It may spend up to an hour in displacing the croc whereupon the other crocodiles usually move away from the vicinity of the hippos. Large crocodiles more than 350 cm (about 11.5 ft.) in length are generally avoided but even these eventually retreat and join the displaced crocs.

Modha (1968) also reports dominance by a hippo on Central Island in Lake Rudolph (now Lake Turkana) in Kenya. There was only one hippo on the island but it had absolute right of way over basking crocodiles, which moved aside on its approach. Only females brooding their nests were seen to snap at the hippo. Similar dominance of hippos over crocs in Uganda and Zambia is described by Cott (1961).

Hippos are occasionally killed by elephants although this interaction is hardly an example of predation. The most usual cause of violent death is an attack by another hippo. Calves are more at risk and are sometimes taken by hyaenas as well as lions but the mother usually manages to protect her offspring adequately.

Being so much smaller, the pygmy hippo is more vulnerable to predation but fortunately for it, lions do not share its range. Leopards do, and Hentschel (1990) provides a photograph of a juvenile hippo that had been killed by a leopard. Other predators of juveniles mentioned by Hentschel include pythons and Nile crocodiles. Very small juveniles and newborn calves are said to be at risk from the golden cat, ratel (honey badger) and civet.

Commensalism

The association between hippos and birds has often been cited as an example of commensalism (Pooley, 1967). One suggested mutual benefit is that the birds use the backs of hippos as resting spots or as look-outs. Pooley observed several species fishing from the backs of hippos including goliath heron, little egret and pied kingfisher. The jacana or lily-trotter was seen to pull leeches from the neck and ears of hippos. In return the hippos are said to benefit from the alarm calls of the birds. Hippos seem to feel insecure on land and make for the water at the slightest disturbance and although they usually have nothing to fear from whatever frightened the birds, they presumably take no chances, e.g. a group of basking hippos stampeded into the water when a flock of white-faced whistling ducks sounded an alarm in response to a swooping fish eagle.

Birds of several species follow hippos as they wade through shallow, weed-clogged water. Some of these are thought to be hunting for coprophagic fish attracted by the hippo's habit of defaecating in the water. Those recorded by Pooley were goliath heron, great white egret, yellow-billed egret, little egret, yellow-billed (wood) stork and spoonbill. Others, such as the Egyptian goose, spurwing goose, white-faced whistling duck and little grebe, are after plant material disturbed by the hippo's passage.

A comprehensive survey of the feeding associations between birds and mammals has been provided by Dean & MacDonald (1981), who divided them into five main categories; the one into which most bird/hippo associations fall is that in which the bird feeds on prey attracted to or associated with the mammal, with or without the bird perching on the mammal. Eight bird species are included with the hippo in this category of which seven perch on the back of the hippo. Five; long-tailed cormorant, little egret, goliath heron, giant kingfisher and pied kingfisher, are known to feed on prey attracted to or associated with the mammal but the feeding association is uncertain in the other two, common cormorant and snake bird (African darter). The eighth, little grebe, as pointed out by Pooley, has a feeding association with the hippo but does not perch on it. Verheyen (1954) saw little egrets and Egyptian geese on the backs of hippos, as have probably most visitors to national parks where hippos exist.

An example of commensalism that spills over into parasitism was described by Ruggiero (1996) between a bird and hippos in Lake Gata in the north of the Central African Republic. It concerned the African jacana or lily trotter, which were seen to run up and down the backs of hippos taking insects and pulling at what were probably ticks or leeches. This is beneficial to the hippo but the jacana's habit of pecking flesh from wounds on the hippos' backs qualifies as parasitism. The hippos bore many wounds as the close confinement in the drying pools led to increased fighting. Several birds congregated at a single wound, which usually bled profusely. Their activities clearly annoyed the hippos, which sometimes vented their feelings on a neighbour, leading to more fighting and further wounding. The hippos attempted to dislodge the birds and sometimes rolled over so that the injuries were submerged but they then came under attack from fish such as *Polypterus africanus*, *Clarius* sp. and *Heterobranchus* sp. The relationship between the hippos and the fish is also parasitic although it is likely that the fish can be beneficial at times by taking leeches and other ectoparasites.

Other examples of birds pecking wounds on hippos include attacks by yellow-billed oxpeckers in Zambia (Atwell, 1966) and one by a common sandpiper, again in Zambia (Gregory, 1985). In the latter case, both red- and yellow-billed oxpeckers were also on the back of the hippo but were not pecking at the wounds and neither was a jacana. On the other hand, Olivier & Laurie (1974b) saw both species of oxpecker tearing flesh from hippo wounds in Kenya. Gregory noted that the sandpiper was also pecking at undamaged skin, presumably for insects, and he wonders if the primary purpose for pecking at the wound was to obtain maggots. The hippo, incidentally, was entirely unconcerned with the activities of the bird, which spent a good 15 minutes probing the open sores.

The opening up of sores on the backs of mammals by birds is quite common. I have seen it at Whipsnade Zoo in England, where crows are a nuisance to white rhinos by keeping open lesions on their backs.

Another animal which I have often seen resting on the backs of hippos is the freshwater turtle. Its presence seems to be neutral as it does not benefit the hippo in any way. I suppose it gains something from the hippo, which makes a convenient mound for basking.

CHAPTER 9

HIPPOS AND MAN

Common hippo in a zoo

RELATIONS WITH MAN

The relationship between human beings and hippos is not always peaceful. The hippo has been hunted from the days when the first men ventured from the forest, either for meat or, more recently, for its tusks, which have a trophy value and which can be carved into *objets d'art*. The prohibition on the sale of elephant ivory in 1989, as a result of placing the African elephant on Appendix I of the CITES[1] regulations, led to fears that the ivory hunters would turn their attention to hippo tusks as a substitute. To some extent these fears have been realised for the amount of hippo teeth in trade has increased, as will be seen below.

Nowadays, hippos killed for any reason are mostly shot but the old system of harpooning is still widely practised. The harpoon carries a float which indicates where the

[1] Convention on International Trade in Endangered Species of Wild Flora and Fauna.

hippo is submerged. Traditionally harpooning is carried out from a flimsy canoe, which can easily be upset by an enraged, wounded hippo. Retaliation by hippos from such attacks is probably responsible for many human casualties. Green (1997) reports being overturned in his canoe by an irate hippo as well as being chased up a tree by another while on land. He maintains that there are several instances a year in Benin of people being injured or killed by hippos and there are plenty of anecdotal accounts elsewhere of human fatalities caused by hippos. There was a mission hospital near to where I used to work in Uganda and one of the commonest injuries treated was loss of part or all of an arm by a fisherman from a hippo attack. Of course, the attack was always unprovoked according to the victim but one wondered how many were due to illegal attempts at poaching a hippo. Hippos do make unprovoked charges, however, whether on land or in the water. I used to spend time boating among hippos when studying fish eagles in Uganda and more than once have been chased by a porpoising hippo. With an outboard engine there was no danger but I would not have liked to rely on paddle power to get out of trouble.

Attacks on land are more difficult to avoid and there are many examples of battered vehicles to prove it. A notorious wreck was the old Landrover used to hunt hippos during the culling operations in Uganda. This boasted several neat holes in its body work where a hippo's jaws had clamped round it. I once met a hippo head-on at night in Queen Elizabeth Park, Uganda. It appeared out of the darkness and hammered the front bumper, to which a radio antenna was attached with steel bolts. The bolts were bent back by the force of the collision, demonstrating the tremendous power generated by a ton and a half of fast-moving hippo but the encounter was not completely one-sided as there were blood stains on the bolts. On another occasion, in daylight, I was subjected to an unprovoked charge from a hippo as I drove past its wallow. Glancing in my driving mirror as I tried to escape, all I could see was the inside of the hippo's throat shortly followed by its muzzle as its jaws closed over my rear light, which was sliced off as neatly as if chopped off with an axe. Although these attacks appeared to be unprovoked, the hippos concerned probably considered the approaching vehicle to be a threat. I suspect that they were males that had been beaten up in territorial fights and that were feeling at odds with the world.

To some people, such as the Valley Bisa of Zambia, the hippo is a "totem" animal and is not hunted (Marks, 1976). At least it was not in the old days but, as with so much of modern life everywhere, traditions are tending to die out. For the most part, however, hippos have been hunted ever since the days when people first came into contact with them.

There is indirect competition, in addition to the direct conflict between man and hippos, resulting from crop raiding that occurs in places where agriculture and hippos co-exist. Hippos do not eat many crops but they can ruin a harvest simply by walking through a field. Some crops, such as rice, are eaten and the losses to the individual farmer can be serious. This aspect is considered in more detail later.

Useful products from hippos

The principal motive for hunting the hippo has been for meat but more recently, the teeth have been valued. The incisors and canines both have a trophy value and can be made into carvings or incorporated as handles into a variety of utilitarian objects such as paper knives or bottle openers. Carved canines are often difficult to distinguish from small elephant tusks at first glance. Other hippo parts that are utilised include the skin, bones, and feet, which can be manufactured into rather tacky souvenirs such as lamp-holders or ash trays. As with many other species, bits of the animal are also used in traditional medicine.

Hippo meat is generally considered to be excellent and is readily eaten over most of the hippo's range. There was certainly no difficulty in finding a ready market for the hippos killed in the Uganda culling operation (see below). Despite the hippo's plump appearance, the meat is not fatty and as with that of so many game species, tastes rather like lean beef. The hippo did not figure in a recent list drawn up and sent to me by Mr Olivier Bilala of animals exploited for food in Gabon. This is not surprising, given that 85% of the country is covered by tropical forest, and it seems that the hippo is not important as bush meat in much of west Africa because of its rarity.

The skin is very thick but the outer dermal layer is quite thin. Most of the hide consists of almost pure collagen, which when oiled, becomes translucent and slices of it can be used as windows. It can also serve as a shield when dried. These traditional uses have rather declined with the coming of manufactured goods but traditional medicine still thrives in many parts of Africa and fuels the demand for hippo carcases. A possible modern use of hippo skins is as shoe leather. When the hippo culling was proposed in Uganda in the 1950s, a leading shoe manufacturer showed interest but the amount of hide that would become available was insufficient to justify the cost of rejigging machines to adapt to the new material.

Bones are collected along with those from other species of mammals and used as fertiliser. Kenyi (1979) reports one such commercial operation in the Queen Elizabeth National Park, Uganda, where bones from 115 hippos were collected from two regions totalling 20.6 km^2. This high density reflects the extensive poaching at the time rather than high natural mortality. The exercise proved too expensive to be worthwhile and was abandoned. It is not, in any case, good practice in a national park as it would result in an unacceptable loss of minerals if carried out on a large scale.

The trade in hippo teeth

The prohibition on the sale of elephant ivory in 1989 seems to have led to an increase in the number of hippo teeth in trade (Weiler *et al.*, 1994). Fig. 9.1 shows the extent of this trade before and after the ban on elephant ivory. The data have been derived by Weiler *et al.* partly from CITES reports and partly from annual customs statistics for Japan and Hong Kong, countries which are known to be important importers of elephant ivory and which conveniently list "all other unworked ivory" in their economic returns. This category excludes waste elephant ivory and refers mainly to hippo teeth. Other forms of "ivory", such as warthog tusks, represent an insignificant proportion of the trade.

Fig. 9.1 shows that the amount of hippo teeth in trade prior to 1989 fluctuated between 2000–4000 kg a year. The small amount traded in 1987 is a puzzle but probably represents under-reporting. There was a noticeable jump in 1989 to nearly 5000 kg (this was the year when export of elephant ivory was banned) and the rise continued in the following year with a peak of 13 681 kg in 1990. Even when allowances are made for the possibility of inaccurate reporting, there seems to be little doubt that there was a massive

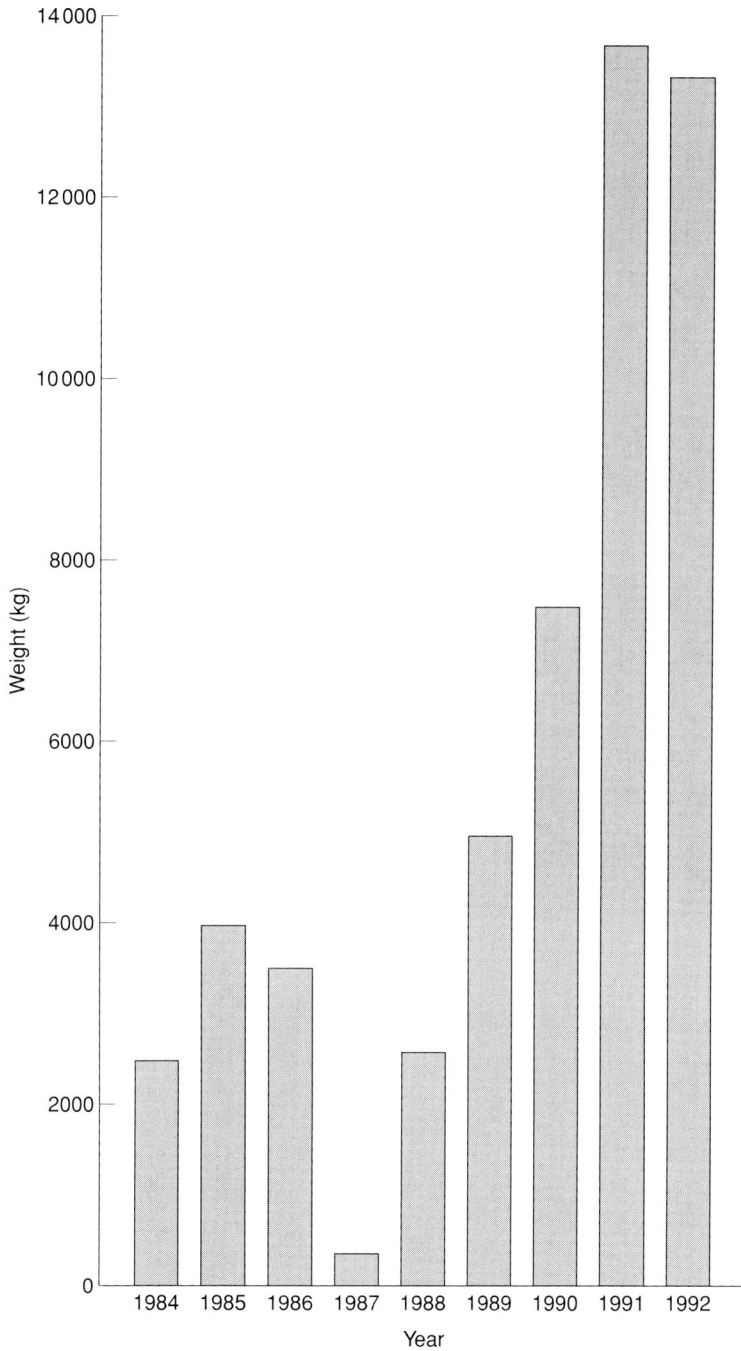

Figure 9.1 Amount of hippo teeth in trade before and after the ban on elephant ivory sales in 1989. After Weiler *et al.* (1994).

rise in the export of hippo teeth from Africa after 1989 and it is difficult to avoid the conclusion that traders in elephant ivory turned to hippo teeth as a substitute.

The above figures refer to the legal trade in hippo teeth but there is evidence of a widespread illegal trade. Weiler *et al.* (1994) quote some examples of consignments of teeth being confiscated by customs because of irregularities in documentation. These include 5 tonnes from Burundi in April 1991 and 10 000 teeth from Zaire in 1993, the latter probably from the Virunga National Park, where many hippos are known to have been shot by soldiers during the recent troubles there. Shooting hippos with modern firearms is easy and carries little risk as they can be killed when bunched together in wallows. It would not be difficult, therefore, to reduce the populations of hippos to dangerously low levels if a lucrative market developed for their teeth.

MANAGEMENT OF HIPPO POPULATIONS

Culling in Uganda

The first large-scale hippo management plan was that established in the Queen Elizabeth National Park, Uganda, in the 1950s. Although the carcases were utilised and sold at a profit, the motive for the project was ecological, not commercial. The cull began in April 1958 and was conducted on an experimental basis in two regions of the park. In one, Mweya Peninsula, a 100% reduction was attempted and in the other, the Kikorongo-Katunguru area of inland wallows to the west of Lake George, only 50% were removed. Half of the wallows area was cleared of hippos and the other half left alone. Hippos from the undisturbed region were subsequently allowed to graze in the cleared area so that for the park as a whole three treatments were applied, namely, a complete elimination of hippos, a 50% reduction, and no disturbance, the latter acting as a control. Any hippos that returned to Mweya Peninsula, the 100% clearance area, were immediately shot.

At first, the hippos were hunted at night, when they were ashore, and were shot from vehicles. The carcases were left where the hippos had been dropped and were recovered the following day. This practice was abandoned after a time for practical reasons. Sometimes the culling crew were not sure of their position when a hippo was shot and they were unable to locate the carcase in the morning. At other times the carcase was found but not until after it had been mauled by scavengers. Night shooting also involved much driving about the park, which added to the costs. Consequently, the cropping team switched to shooting hippos in the water by day, either from the shore or from a boat. This was easy since the animals were crowded together in a relatively small area. When a hippo is shot in the water, it immediately sinks but after a few hours, fermentation of the stomach contents produces gases that inflate the carcase and float it to the surface from where it can be towed to the bank and hauled ashore with the aid of a tractor.

Once ashore the carcase was processed by the culling team. Hippo meat is highly palatable and is much sought after by local people. The park authorities did not want to waste the meat and at first it was given away but later, when it became apparent that there was going to be a lot of meat to dispose of, each carcase was sold for a set fee. The parks did not wish to become involved in the retail meat trade so responsibility for the disposal of the shot hippos was passed to an agent, who made arrangements with local butchers to bring their vehicles into the park and collect a carcase. Orders were taken beforehand and the butcher was told where and when he should bring his truck to collect his hippo. He was then free to sell it for whatever profit he could make in the neighbouring villages. The teeth were retained by the parks for later sale to the carving industry as a form of "ivory". As mentioned earlier, some interest was expressed in utilising the skin for shoe leather but the quantity potentially available was insufficient for it to be a practical proposition.

Many of the carcases were subjected to scientific investigation and much of the information given in this book, particularly on reproduction, derives from these studies. In the beginning, the last of the Fulbrighters, Bill Longhurst, and some local scientists processed the carcases but it was obvious that a more detailed analysis was required and negotiations were instituted to establish a long-term scientific presence in the park. These terminated in the founding of the Nuffield Unit of Tropical Animal Ecology in 1961, a few years after the culling began. The Unit, funded by the Nuffield Foundation and affiliated to Cambridge University, did not have to look far for its first research project. The Director, Dick Laws, extended the culling area to include that part of the eastern shore of Lake Edward that was centred on Lion Bay.

The hippo culling in Uganda was the first to be carried out on such a large scale. It was also one of the best that has ever been attempted as it satisfied many of the criteria usually considered desirable for the effective management of wildlife. First, it achieved the desired management function in preventing the possible collapse of the park's ecosystem. Secondly, it provided the local people with a source of food or income in return for their toleration of the animals in their region. Thirdly, the park itself made a modest profit so that the programme was self-funding. Finally, the management action was also an experiment and a wealth of scientific data was accumulated which could be used for the more effective management of the park. In many ways the project was ahead of its time and it provided a model for others to follow. Some of these were commercial but others were carried out for management reasons although, as in Uganda, the carcases were not wasted.

Culling and cropping elsewhere in Africa

Some mention of hippo culling in other African countries has been made in Chapter 5 where reproduction was considered but for completeness they are summarised here.

Tanzania

A cull of some 300 hippos was carried out in the Rufiji Delta in Tanzania between April and June of 1994 (K. Leggett, *in litt.*). Further culling was proposed for 1995 but I have not been able to find out whether or not it took place. The motive for the culling appears to have been commercial but the opportunity was offered to scientists to carry out biological studies of the carcases. Unfortunately, insufficient notice was given in 1994 for the offer to be taken up and I have not come across any publications stemming from this or subsequent culls.

South Africa

Culling of hippos in the Kruger National Park has taken place on several occasions in response to habitat damage and perceived loss of condition due to drought. The first cull was carried out in the Letaba River during the drought of 1964 when 104 hippos were shot (Pienaar *et al.*, 1966). Further mortality occurred during the peak of the drought in 1970/71 when at least 150 hippos died in the Letaba and Olifont Rivers from exposure and malnutrition. The crowding of the animals into the few remaining wallows led to severe fighting and more deaths. To prevent a reoccurrence, it was decided to maintain the population at the reduced level reached at the end of the drought. Consequently, 225 hippos were shot from May to September, 1974, and in August 1975. A further 238 were taken out during 1976 and 1977. The culls were made in the Crocodile, Lebata and Olifants Rivers and in Orpen Dam (Smuts & Whyte, 1981).

Further culling in the Kruger National Park took place ten years later in the Sabie River, where 100 and 68 animals were taken out in 1987 and 1988 respectively (Viljoen & Biggs, 1998). An additional 26 were culled in nearby wallows and 10 were removed for translocation to other game reserves. Culling does not necessarily involve the destruction of animals if the sole objective is to lower their density and such live capture is as effective as shooting.

Zambia

The culling in Zambia during the early 1970s differed in that the carcases were brought to an abattoir for processing rather than being butchered in the bush as in the Ugandan situation (Marshall & Sayer, 1976). This was presumably the abattoir built to process the elephants that had been culled in the Luangwa Valley (Hanks, 1979). The construction of an abattoir puts the cost of the operation at a level that might detract from the economic viability of commercial cropping but if one of the objectives is to produce meat for human consumption, the procedure must meet strict hygiene standards. These require an abundant supply of water for cleaning the carcases and disposing of the offal, and the Zambians were fortunate in having a large river close to hand. Hygiene was not the primary consideration in the Ugandan culling although conditions at the culling sites were probably no more unhygienic than those at domestic butcheries in the area.

Further culling of hippo was carried out in the Luangwa Valley region in 1995 and 1996, apparently for commercial reasons although numbers were thought to be at their maximum carrying capacity. The culling was carried out by Hawk Trading, a Zambian company, and the meat sold but the carcases were made available to scientists for biological examination (K. Suzuki, *in litt.*). The results of the analyses do not seem to have been published but some unpublished reports were kindly sent to me by Mr Suzuki. These show that in 1995, 507 hippos were shot in 32 hunting days between 25th May and 15th July. In the following year, 234 were shot in 22 hunting days between 21st August and 15th September. Details of the scientific data collected have been incorporated into the relevant chapters of this book but no studies were carried out on the effects on the environment of removing these animals. When these reports were written, sufficient time would not have elapsed for any changes to have become apparent but in any case, the numbers involved were rather too low in proportion to those present for a significant effect to be expected.

Zimbabwe

Some culling took place in the Gonarezhou National Park in 1972 and 1978 when 160 and 105 animals respectively were shot in response to perceived overgrazing

(O'Connor & Campbell, 1986). Despite this, numbers continued to rise, from immigration as much as from reproductive increase.

Practical considerations in the exploitation of hippos

The technique for "harvesting" hippos is fairly well standardised and follows closely the practice adopted by the Ugandan hunters in that hippos are shot in the water and towed ashore when fermentation of the gut contents produces enough gas to bring the shot animal to the surface.

The offtake of hippos in a cropping scheme would need to be sustainable and, for economic reasons, it would be best to take the maximum sustainable yield (MSY) but the latter is extremely difficult to estimate. There are several mathematical ways of doing so, all of which are fraught with danger because if the MSY were exceeded, the population would quickly slide to extinction. No cropping scheme has been carried out long enough to put the matter to the test but one calculation of the MSY was made by Ndhlovu & Balakrishnan (1991) for a hippo population in the Luangwa Valley, Zambia. This was based on a formula applied to all large mammals in the community and would have involved taking 9% of the hippo population each year but as far as I know, the attempt was not made. Perhaps this was just as well as the offtake seems rather high.

The processing of meat for sale has been a problem in many game cropping schemes because the product is heavy and perishable but not very valuable. Consequently, transport costs represent a large proportion of the expenses and reduce the profits. By the nature of wildlife distribution, markets tend to be some distance from the cropping zones so that the meat has to be transported over long distances. The choice of a permanent or mobile abattoir and the provision of electricity are among the factors to be considered. These and other complications have been considered in my book on wildlife utilisation (Eltringham, 1984) in which I showed that many cropping schemes were profitable only through the sale of skins and other trophies. These have the advantage of being valuable and not immediately perishable as well as requiring only the minimum of processing in the field. Hippo cropping fulfils these conditions as the teeth and skin have commercial value but the recruitment rate of the species is not high and there would be a serious risk of over-exploitation if cropping were to be carried out on a sufficient scale to be attractive to entrepreneurs.

Kingdon (1997) maintains that the hippo has outstanding potential for domestication or ranching but the start-up costs would be high and although perfectly feasible, I doubt that domestication will ever happen. Nevertheless, hippos tame very quickly, even as adults, and those reared in captivity are usually docile. A colleague of mine, Dr Michael Lock, sent me an amusing cutting from *The Hants and Dorset Avon Advertiser* dated, tellingly, 1st April 1998, which showed a photograph of two hippos from Marwell Zoo being used as lawn mowers in the Close of Salisbury Cathedral. They also doubled as muck-spreaders. Some of the droppings, packaged in small hermetically sealed bags for sale in the cathedral's gift shop, were said to have sold like hot cakes, particularly to Japanese tourists! Unfortunately it was an April Fool joke but the idea has its attractions.

CROP-RAIDING BY HIPPOS

Crop-raiding by hippos occurs throughout Africa but it tends to be a problem only where large hippo populations occur next to dense human settlements. In crowded west Africa there are too few hippos to cause widespread damage to crops although losses can be severe on a local scale. Many of the big populations of hippos in east and southern Africa are confined to large national parks and again, the possibility of hippos encroaching onto cultivated land is limited. Where agriculture extends to the borders of such parks, however, conflict can occur. Damage is caused in two ways: first by the more obvious eating of crops and secondly by trampling. Rice is likely to be taken by hippos as the plant is a graminid and closely related to the natural grass food of the hippo, as are the cereal crops. Trampling, on the other hand, is not limited to any particular crop.

Surveys of hippo damage have been made in Malawi, a country with a high density of both people and hippos. Not surprisingly the greatest number of complaints originated in regions where human populations were densest (Mkanda, 1994). Mkanda & Kumchedwa (1997) investigated the possibility that farmers exaggerated crop losses and decided that they did as far as grazing was concerned but underestimated damage caused by trampling. They suggested that trampling was tolerated more because the trampled crops were likely to recover before harvest. Maize was the main crop eaten by hippos in Malawi, particularly early in the rainy season. Rice was taken to a much lesser extent and sugar hardly at all. To some extent this reflected the relative proportions in which the crops were grown.

Whether or not hippos were perceived as pests was one of the queries included in the questionnaire distributed to participants in the survey to determine the numbers and distribution of hippos described in Chapters 10–12. There were two main complaints against the hippo: one that it attacked fishermen and the other that it damaged crops. Not all respondents answered the question, presumably because they had no information, but of the answers received there were 19 positive statements that crop-raiding was a problem and 11 that it was not. Some of the latter may have indicated that the correspondent had not heard of any damage for there were one or two countries from which contradictory reports were received.

Complaints were not confined to any particular region of Africa although, not surprisingly, they tended to originate mainly in countries with large hippo populations. Zambia, which holds the largest number of hippos in Africa, was also the country with most complaints, all of which indicated that the problem was severe. There were almost as many complaints from Burundi, a small country with only an estimated 1000 hippos. It is, however, densely populated and consequently the chances of hippo/human contact are high.

Residents in The Gambia sought advice from the IUCN[2] Hippo Specialist Group and were advised to try physical barriers including electric fencing, which proved to be quite effective, at least in the short term. The area concerned, however, was small, as was the amount of fencing required, and such barriers might be too expensive to install on a larger scale. Experience from Lake Naivasha in Kenya suggests that the electric fences are most effective if they are rendered conspicuous, e.g. by attaching

---

[2] International Union for the Conservation of Nature and Natural Resources.

ribbons to them. It seems that the hippos will then avoid a close approach, possibly because they have previously experienced a shock, whereas if they simply blunder into the fence, they are as likely to rush forward as to retreat, destroying the fence in the process. Hippos could, of course, be kept out by more substantial fencing coupled with ditches but apart from the construction costs, maintenance would be prohibitively expensive. It is not only money that is required to maintain fences, for enthusiasm is also important and that is a factor that is apt to wane once the immediate danger has passed.

Active defence by the farmer, involving shouting and banging of tins etc., is the most widely practised response to marauding hippos in Malawi although farmers much prefer them to be shot (Mkanda & Kumchedwa, 1997). The latter has to be carried out by the authorities and not by the individual farmer, and not all requests for assistance are met, e.g. only 147 of 2167 calls were answered in the period 1984 to 1989. Barriers are not used to any great extent. Attempts to drive hippos away are never wholly successful and people run the risk of being killed as a consequence.

It has to be admitted that crop-raiding by hippos is an unsolved problem. Mkanda (1994) considered the effectiveness of shooting as a means of controlling marauding hippos in Malawi and came to the conclusion that it was ineffective in preventing damage. Over a six-year period, 1984–1989, a total of 928 hippos were shot and 651 wounded. Later figures given by Mkanda & Kumchedwa (1997) for one region, Elephant Marsh, in Malawi were 354 killed and 255 wounded between 1991 and 1994. These kills seemed to have had no effect on the level of crop damage. The high proportion of woundings gives cause for concern from an animal welfare point of view.

No attempt has been made, continent wide, to assess the total cost to agriculture from depredation by hippos and it may be that it is unimportant, but it is nevertheless of great significance to individual farmers. For example, Mkanda & Kumchedwa (1997) estimated that, in 1993–94, 22.2% of the maize and rice crop was lost to hippos on 65 ha of farmland in Malawi. Human populations are increasing and grasslands are becoming converted to agriculture, two factors that could spell disaster for hippos. Various solutions are possible, such as compensating farmers for losses, and there is need for urgent research into the matter.

Fortunately, the pygmy hippo does not appear to cause any serious damage to crops, and risks to its survival are different.

TRANSLOCATION OF HIPPOS

One solution to crop raiding is to shoot the hippos but another is to capture them alive and translocate them to another place. This can work if there is somewhere for them to go to. Catching and moving hippos is not a task to be undertaken lightly and it is not surprising that it has rarely been attempted. One example I know of is that reported in *African Wildlife*, the magazine of the Wildlife and Environment Society of Southern Africa (Anon, 1993). The hippos, said to be a breeding nucleus, were released by the South African National Parks Board into the Sundays River area of the Zuurberg National Park in the Eastern Cape, where they once existed although none had been sighted there since 1860. Their place of origin is not mentioned so one doesn't know how far they had to be transported. During a recent visit to South

Africa, I saw a group of hippos in the Pilanesburg National Park which I was told had been introduced from the Kruger.

HIPPOS IN ZOOS

Hippos make good zoo animals. Both species breed well, although there is room for improvement, and their potential tameness makes them easy to handle. There should be no need to capture further specimens from the wild unless the introduction of new blood is considered desirable. Most zoo specimens are, in fact, captive-bred. Of the 99 common hippos recorded by the International Species Inventory System (ISIS) in an incomplete census made in 1985, 68% were captive-born. The breeding groups are small and scattered, however, and the reproductive rate is below that which would occur in the wild, e.g. only eight calves had been born in the previous year in the above sample and, of these, three died within a month of birth. Zoos are not necessarily trying to produce as many hippos as possible, for being so large, hippos are expensive to accommodate and feed.

The situation with pygmy hippos is much the same. The latest issue of the *International Studbook for the Pygmy Hippopotamus* (Tobler, 1988) lists 820 animals of which most if not all had been born in captivity. Not all these hippos were still alive but at the time of publication, there were 340 living animals (131 males, 206 females and three of unknown sex). The total number of pygmy hippos in zoos has more than doubled over the past 25 years and it is ironical that the future of the species seems more assured in captivity than in the wild. As with the large hippo, however, zoos are not breeding pygmy hippos to their full potential. Mortality in young calves is too high and many deaths are caused by maternal neglect and by injuries inflicted by the mother. This suggests that husbandry techniques are less than ideal. A potential problem is the widespread practice of keeping pygmy hippos as monogamous pairs, which is unlikely to be the natural situation and the undivided attention of the male may well be stressful for the female. The studbook mentions attacks by the mate as a cause of death. There is also an apparent assumption that the pygmy hippo is mainly an aquatic creature and although a large pool is usually provided, less attention is given to the animal's terrestrial needs.

THE CONSERVATION OF HIPPOS

The conservation of hippos is touched upon in the chapters (10–12) concerning the numbers and distribution of the species and the subject has been considered in some detail in the IUCN Action Plan for hippos (Eltringham, 1993c). Some of the recommendations in the Action Plan are counsels of perfection and others are statements of the obvious but the publication has the benefit of comprehensiveness. The stated objectives of the Plan were:

1) To establish the distribution of the two species, particularly in those range states for which information is sketchy or non-existent at present.
2) To identify populations that are at risk.

3) To maintain populations of both species within their present ranges.
4) To disseminate information about the conservation status of hippopotamus.
5) To investigate ways in which to exploit the commercial value of the common hippopotamus.
6) To encourage zoos to co-operate in building up and maintaining viable breeding groups in captivity.

Each species will be considered in turn to see how far these objectives are being met.

Hexaprotodon liberiensis

The biggest threat to the pygmy hippo in the wild is deforestation and although the threat is not imminent, provision should be made to ensure that suitable forests are conserved to provide adequate habitat. An area of secure forest, which contains at least some pygmy hippos, lies in Sapo National Park in southern Liberia and conservation efforts should be concentrated in that region. It would be unwise to rely solely on this park and attempts should be made to identify other suitable areas and to ensure that they are given protection. The other main threat is hunting although it is not clear how serious this threat is to the survival of the species. Political unrest in Liberia has made it almost impossible to attempt any conservation work for many years but the situation is now calm and such activity is once again feasible (P.T. Robinson, *in litt.*). Much more needs to be known about the numbers and distribution of the pygmy hippo throughout its range and reliable techniques for counting animals need to be developed. These may have to be indirect, such as counting dung deposits, which has proved to be successful with forest elephants.

The distribution of the species outside Liberia should be established so that conservation action may be planned. In particular, further attempts should be made to establish the status of the pygmy hippo in Nigeria especially as it is a distinct subspecies. There is probably little point in investigating the reported existence of the pygmy hippo in Guinea Bissau as the single sighting there was almost certainly of a small common hippo.

The position of the pygmy hippo in captivity needs to be considered in view of the poor breeding record of zoo specimens. Experiments with alternative husbandry methods should be instituted, with special attention being given to the provision of an adequately sized run with plenty of cover in order to simulate the natural habitat of the creature.

Hippopotamus amphibius

The large hippopotamus needs a secure aquatic refuge by day and adequate grazing grounds by night. Shortage of water is unlikely to be a great problem although the security of hippos in the water is another matter. Loss of grazing grounds to agriculture has occurred in a number of places and may well be the biggest risk to survival. As a species, the hippo is not in any immediate danger of extinction. It is, however, potentially vulnerable because of its specialised habits. A simple change of course of a river can have serious repercussions on isolated populations and the drying up of water masses can be disastrous, for the resident hippos seem to have a limited ability to disperse. I once saw a group of a dozen or so hippos near Maun in Botswana trapped in a large wallow that was drying up. The nearest water was a long way away with human settlement in between. I thought it was extremely unlikely that they would survive. They could possibly have been saved by artificial feeding and pumping in water but in an area where people and their livestock were suffering from severe drought, that was not really an option. The local people were showing great interest in the hippos and it is significant that none of them was harassing them in any way. One rather got the feeling that they were sympathetic towards the hippos' plight.

The greatest risk to hippos is undoubtedly in West Africa. Not only is the total number low – about 8000 – but the hippos are scattered over a wide area in small groups, often under 50, which is well below the number that geneticists tell us is necessary to ensure the long-term survival of a population. The first priority, therefore, should be to build up numbers in these isolated populations, possibly including the importation of animals from elsewhere in order to increase the genetic diversity. The loss of genes through inbreeding is a serious matter as the population is then less able to respond through natural selection to long-term environmental changes. It is for this reason that even small populations should be conserved for they may be important components of the gene pool with rare alleles that could make all the difference to survival or extinction in the future. West African countries in which the hippo is most at risk have been identified (see Chapter 9) and include Burkina Faso, Cameroon, Central African Republic, The Gambia, Ghana, Ivory Coast, Mali, Nigeria and Sierra Leone. The survival of the west African hippo is important zoologically, for its loss could mean the extinction of a subspecies, *H. a. tschadensis*, if such a taxon exists (see Chapter 1).

The tendency for large numbers of hippos to concentrate in restricted areas makes them particularly easy to poach but by the same token, it also makes them easy to guard. A priority in hippo conservation, therefore, is to identify aggregations that are at risk and to institute adequate policing of them. Although law enforcement will always be necessary, it is not a solution to the problem of illegal killing, for the patrolling rangers cannot be everywhere. There is no doubt that the survival of hippos, as of any other large mammal, depends on the people who have to live and work among them. Hippos are not always welcomed because of the damage they cause to crops and the threat they pose to life and limb. Crop damage has already been discussed and there is no easy solution to the problem. An attempt should be made to assess the degree of financial loss with a view to providing compensation but perhaps a better approach is to avoid the juxtaposition of hippos and agriculture. This is beyond the powers of unofficial conservation bodies to achieve and it could only be done at governmental level.

Confrontation with hippos is not inevitable for we had few problems in the national park in Uganda, where I used to live. There were fishing villages within the park and women could be seen washing clothes in the lake within a couple of hundred metres of wallowing hippos. Where they are used to such human activities, hippos are not aggressive but if they are harassed, it can be a different matter. Fishermen are said to be at particular risk and it would be instructive to investigate the circumstances leading up to such attacks as some at least appear to have been provoked.

Local people are more likely to accept the presence of hippos if they can derive some benefit from them. Meat is an obvious benefit but it is doubtful that there are many places where hippos could be cropped on a sufficient economic scale and on a sustainable basis. The sale of hippo products, such as teeth or skin, would bring in extra income but probably not enough to satisfy the expectations of the expanding human populations within the hippo's range. There is little point in stimulating a demand for wildlife products that cannot be met in sufficient quantity, for that is a road to extinction. Although I feel such pessimism is justified, I do not suggest that research into the possibility of exploiting hippos should not be carried out. Factors in favour include the relatively short gestation and lactation periods of 11 months and 12 weeks respectively. The potential for rapid reproduction, therefore, is high but a cropped population would have to have an initial size of many thousands if the enterprise were to be economically viable.

Game viewing is a benign form of generating income, and hippos, along with other large mammals, are certainly a tourist attraction. A problem with eco-tourism is that very little of the money it brings in finds its way into the pockets of the people who have to co-exist with dangerous and destructive wildlife. A more localised arrangement, similar to the CAMPFIRE project in Zimbabwe, in which residents are involved in decision making, would go a long way to achieving general acceptance of the value of wildlife conservation.

Effective conservation of a species requires a detailed knowledge of its biology and much remains to be found out about the hippo. My 1993 survey, defective though it was, is the only attempt so far to assess the total number of hippos in the world. There are many gaps in its coverage. No recent information is available for Angola, Chad or Rwanda, countries which, in the past, contained flourishing populations of the species, and only very incomplete returns were received from Cameroon, Ethiopia and Malawi. Further censuses are clearly necessary and equal attention should be paid to all countries, whether or not they have flourishing populations.

If the day ever arrives when the survival of hippos depends on captive breeding, some improvements over the present regimes in zoos seem desirable. Choosing the appropriate individuals for mating is important but as with similarly sized mammals, transport of adults between zoos around the world is not easy and the technology for artificial insemination or embryo transfer has yet to be developed for hippos. Specialist hippo "farms" would seem to be the best approach, perhaps combined with domestication as proposed by Kingdon (1997) for unless the enterprise was revenue-earning, it would be unlikely to materialise.

CONCLUSIONS

The fourth objective of the Hippo Action Plan mentioned earlier – the dissemination of information about the conservation status of hippos – is one that was not mentioned in the above discussion because very little has been done in this direction. There are some individuals, other than scientists, working towards hippo conservation and in the course of writing this book, I came across a small organisation involved in the conservation of hippos in Zimbabwe (Turgwe Hippo Trust, P.O. Box 322, Chiredzi) and there may be more. It is certainly an area in which amateurs can make a significant contribution to conservation. Much remains to be found out about both species and I hope that this book will stimulate more people to take an interest in this unique and fascinating family of animals.

CHAPTER 10

THE DISTRIBUTION AND NUMBERS OF HIPPOS

The two hippo species with their distributions

GENERAL INTRODUCTION

L iving hippos are now confined to Africa although the distribution was much wider in the Pleistocene. Some species, now extinct, occurred in Europe including Britain, particularly in the region of England now covered by the fens, where the remains of many specimens have been discovered. Details have been given in Chapter 3 in which the prehistory of the group was considered. Of the two living species, the large hippo has by far the widest distribution for the pygmy hippo is now more or less confined to Liberia although there are reports of sizeable numbers in the immediately adjacent countries, especially the Ivory Coast.

The general distribution of the large hippo has been well known for centuries and it is probably true to say that the species occurred anywhere throughout sub-Saharan Africa wherever conditions were suitable, i.e. wherever there was water and grazing.

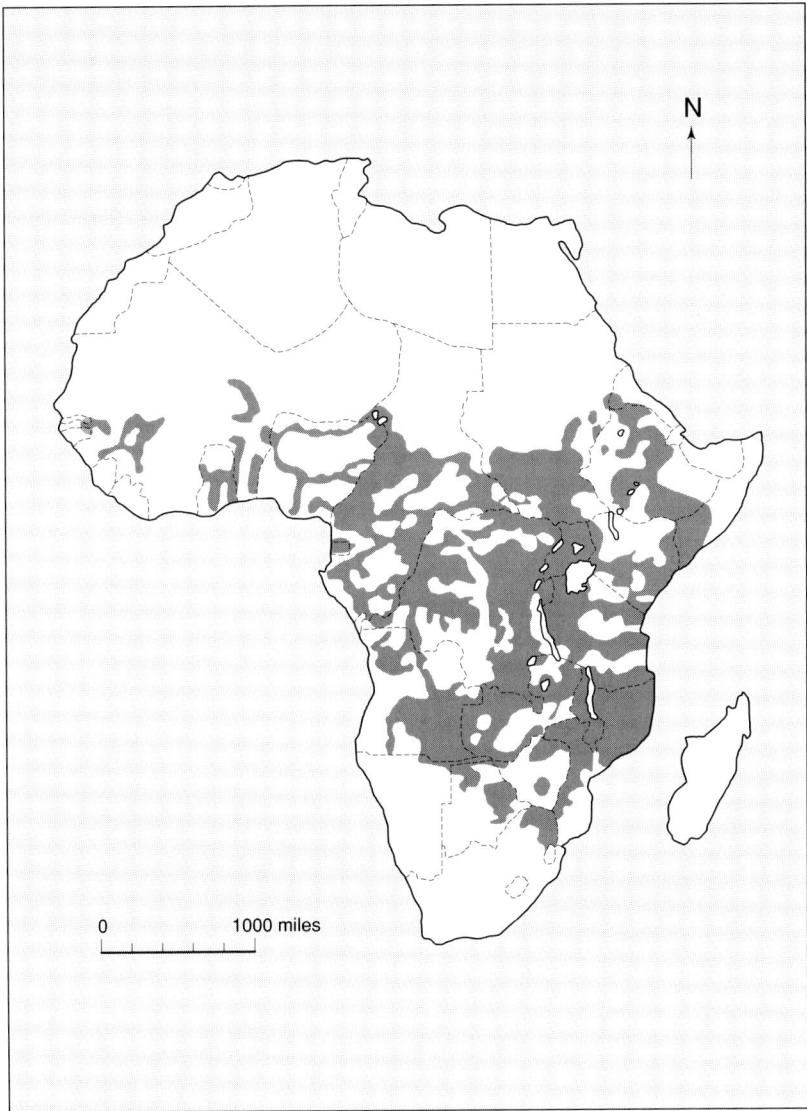

Figure 10.1 Distribution of the common hippopotamus around 1959 according to Sidney (1965).

The first serious attempt to plot its distribution was made by Sidney (1965) in her survey of some of the ungulates of Africa. Her map of the hippo's distribution is reproduced as Fig. 10.1, which shows the species almost everywhere including unlikely forested areas. A more realistic map would show the species confined to rivers and lakes. Rivers flowing through thick forests do not contain hippos unless there is some grazing not far away.

In order to find out more about the distribution of hippos, I was given the task of assessing more precisely their numbers in the various countries of Africa. This formed part of the research programme of the Hippo Specialist Group, of which I was then the Chairman. This is one of a number of specialist groups within the Species Survival Commission (SSC) of the World Conservation Union, perhaps better known as IUCN, the International Union for the Conservation of Nature and Natural Resources. The specialist groups are not confined to taxonomic units but also cover "subject" areas such as re-introductions and wildlife veterinarian work. The groups are made up of experts in the various fields who give their services voluntarily. Some groups are large and produce regular newsletters. Others, such as the Hippo Group, with only two species to worry about, are small and act as advisory bodies. The SSC produces a regular newsletter, *Species*, which gives information about the activities of the various specialist groups. An important function of these groups is the production of an action plan, which assesses the current status of the group and draws up plans for the conservation of the species within it. Action plans for taxa which contain a large number of species are necessarily longer than those with only a few so the Hippo Action Plan, covering only two species was combined with that for the Pigs and Peccaries Specialist Group. The joint plan was published in 1993 (Oliver, 1993) and provides the most up-to-date information on the status of the hippo on the continent as a whole.

The Action Plan survey was not only the most complete, it was also the first attempt to count the number of common hippos in the whole of Africa so that it is impossible to say whether or not numbers have increased or decreased recently. At the time, the perceived wisdom was that numbers had declined and this is probably not far wrong but one cannot be sure. What one can sometimes do is to compare serial counts made in restricted areas and such counts do often show a marked decline in numbers, as we shall see when the countries are considered separately. Despite the assumed decline in numbers, there was little obvious signs of concern by conservationists over the fate of the common hippo and it was afforded no special protection. It was not, for example, included in any of the appendices of CITES, the Convention on International Trade in Endangered Species of Wild Fauna and Flora, except for Appendix III by Ghana. CITES, also known as the Washington Convention, controls the trade in wildlife. Species on Appendix I are not allowed to be traded but those on Appendix II may be within certain rigid limits. Appendix III lists those species given local protection by individual countries. It is not used very much and, frankly, is of little practical value to the conservation of species.

The SSC survey was carried out between 1988 and 1989 and was conducted mainly through correspondence augmented by some personal observations. Letters and questionnaires were sent out to residents in the countries concerned and to people with recent knowledge of hippo distribution. Inevitably, the quality of the information so gathered varied very considerably. Some estimates of numbers were no more than educated guesses whereas others were based on detailed counts. Replies were received from a total of 55 correspondents covering 34 countries. These were all the known range states for the species with the exception of Angola, for which no information could be obtained because of the disturbed political conditions existing there at the time. It cannot be claimed that every group of hippos has been counted and even in those countries where the cover was good, it is likely that some populations were overlooked. The results, therefore, provide only minimum totals, although even that conclusion assumes that the counts themselves were reasonably accurate. Hippos are not the easiest animals to count since most, if not all, of the body is submerged in often murky water and it takes a long time watching a school of hippos bobbing up and down before one can get a fair idea of the number present. Even so, a rough idea of the size of a population is better than no idea at all and the results of the survey are probably of the right order of magnitude.

For the purposes of the survey, no notice was taken of the supposed subspecies of *H. amphibius* for even if they are valid taxa, they cannot be distinguished in the field. Consequently, the distribution is based on geographical considerations. For convenience, totals were recorded for individual countries. This is also useful for conservation purposes because the degree of attention paid to conservation and management often differs very considerably from one country to another. Each country is considered here in turn with a standard format. First, the distribution is described, then the numbers are estimated and finally the conservation status is discussed. The terminology used to describe the conservation status follows the conventions used for all SSC Action Plans.

The continent was arbitrarily divided into west, east and southern for the presentation of the results. The remainder of this chapter is concerned with West Africa.

THE DISTRIBUTION AND NUMBERS OF HIPPOS IN WEST AFRICA

Hippos need water but the larger species needs grass as well so it is absent from the rain forests except near large rivers where there is some grazing along the banks. As most of West Africa was originally heavily forested, one would not expect many hippos to be present in this part of Africa and this was indeed found to be the case. There are very few in the central Congo rain forests and most are found in estuarine habitats and the lower reaches of rivers, where human activity has often converted forests into grassy woodlands. Hippos will often venture into the sea at the mouths of rivers, e.g. in the estuary of the Casamance River of Senegal. A population in the Archipelago of Bijagos, off the coast of Guinea Bissau, appears to be fully marine when not grazing.

The number of hippos in each west African country will be considered in turn and this procedure will be followed for the other regions of the continent. References cited as "*in litt.*" refer to information provided by respondents to the survey. It is not always possible to assign a group of hippos to one country rather than to another as rivers are often used as international boundaries. The distribution of the pygmy hippo will be considered here as it occurs only in West Africa.

THE DISTRIBUTION AND NUMBERS OF *HEXAPROTODON LIBERIENSIS*

Fig. 10.2 shows the distribution of the pygmy hippo over its whole range and, in more detail, throughout Liberia. Its distribution in the past was probably similar to its distribution today although the populations are now more fragmented. The bulk of

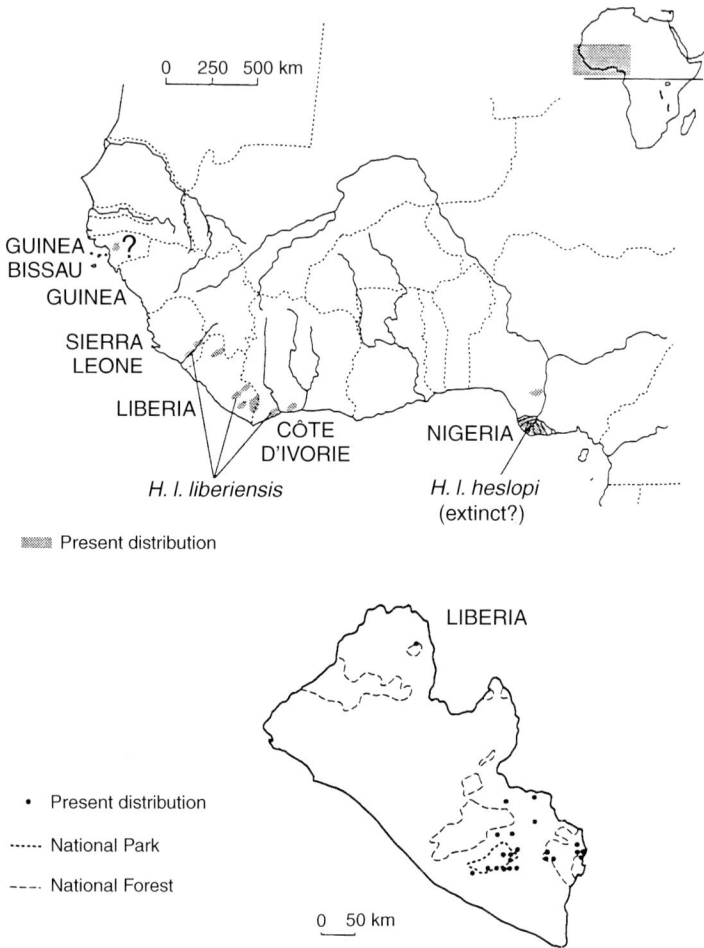

Figure 10.2　The distribution of the pygmy hippopotamus in West Africa with recent records from Liberia. From Eltringham (1993b).

the population is centred in Liberia (Anstey, 1991) but the species also occurs in the three contiguous countries, namely Guinea, Ivory Coast and Sierra Leone.

A smaller population, belonging to a separate subspecies, has been reported from Nigeria, some 1800 km away and separated from the main population by the Dahomey Gap, a break of considerable antiquity in the otherwise continuous rain forest cover along the west African coast. Such a discontinuous distribution on either side of this gap is very rare in forest vertebrates and led Robinson (1970) to doubt the existence of the species in Nigeria. The evidence for its presence, however, is good for Heslop (1945) describes shooting one near Omoko on the Niger Delta. It is possible that it was a young common hippo but the description by Ritchie (1930) of two skulls collected from the Niger Delta in 1928 is convincing and leaves little doubt that the species did occur in Nigeria even if it does not exist there now. The evidence for its continuing presence is equivocal. F.O. Amubode (*in litt.*) assumes that the species does still occur but provides no evidence for his views. It never was very numerous and there is a real possibility that it is now extinct in Nigeria.

Another isolated population has been reported from Guinea Bissau on the Corubal River by Cristino, who claims to have shot one (Cristino & de Melo, 1958). It is more likely, however, that the animal was a small common hippo.

Guinea

Distribution: The pygmy hippo occurs in three rivers bordering Liberia and Sierra Leone (L. Macky, *in litt.*). The species is found in the Reserve de Ziama on the Liberian border in the south of the country.

Numbers: A few individuals occur according to Macky in groups of six or seven. The population is said to be almost static. It seems likely that numbers should be counted in tens rather than hundreds.

Conservation status: Subject to complete protection by law. Macky maintains that the species is not threatened and that its conservation status is satisfactory.

Guinea Bissau

Distribution: Said to occur on the Corubal River but, as mentioned above, the reported specimen was almost certainly a young common hippo and consequently it is safe to assume that the species is absent from Guinea Bissau.

Numbers: Not applicable as the species does not occur in the country.

Conservation status: It is probably not worth the time and money it would take to show that the species is absent and, therefore, of no conservation interest.

Ivory Coast (Côte D'Ivoire)

Distribution: Reported by A. Blom (*in litt.*) as being present in the Fresco region, which contains one of the last areas of coastal forest remaining in the Ivory Coast. The pygmy hippo occurs in the Dagbe, Bolo and Niouniourou Rivers and also around the Tatigbo Lagoon. These rivers flow into the Gulf of Guinea and the stretches thought to hold pygmy hippos are on or near the coast. Tatigbo Lagoon is on the coast itself. The species is also reported from Mount Nimba Reserve, where the borders of Ivory Coast, Guinea and Liberia meet, and in the forests of Tai National Park and the contiguous N'Zo Faunal Reserve in the south-west of the country (IUCN/UNEP, 1987).

Numbers: No estimates of numbers are available.

Conservation status: No recent information is available but the presence of the species within national parks and reserves is reassuring assuming that law enforcement is effective.

Liberia

Distribution: Widely dispersed throughout the country with most recent records from the Sapo National Park (A.L. Peal, *in litt.*). Isolated populations reported elsewhere in the following counties: Cape Mount, Grand Bassa, Grand Gedeh, Lofa, Maryland, Nimba and Sinoe (M.E.J. Gore, *in litt.*).

Numbers: No reliable estimates have been made of the population size, and estimates are largely guesswork. Nevertheless, total numbers are likely to be in the thousands.

Conservation status: Again, no reliable information is available but correspondents in the SSC survey considered the species to be at risk through loss of habitat and to a lesser extent through hunting for food. Some use of the teeth for the fake ivory trade was reported.

Nigeria

Distribution: The distribution of this separate subspecies (*H. l. heslopi*) is in and around the Niger Delta near Port Harcourt. Powell (1993) reviews some past records and lists the following localities where the pygmy hippo has been seen or shot by hunters: Aboh, Akpede, Akunomi, Biseni, Ikibiri, Isemu, Lalagbene, Odi, Omoku and Sabagreia on the Nun River between Kaiama and Yenagoa.

Numbers: Most correspondents consider the subspecies to be extinct although F.O. Amubode (*in litt.*) believes that it still exists. Heslop (1945), who described the subspecies, put the total number at 30 with the largest population numbering seven.

Conservation status: Obviously, if the subspecies is extinct, there is no point in worrying about its conservation status. The most important step in its potential conservation is to ascertain whether or not it still exists.

Sierra Leone

Distribution: The pygmy hippo occurs in the south-east of the country not far from the Liberian border. It is found mostly in the Gola Forest Complex which comprises three separate forests (G. Teleki, *in litt.*) but a population also occurs on Tiwai, an island in the Moa River near Potoru, and another in the Loma Mountains in the north-east.

Numbers: G. Teleka (*in litt.*) reports that his survey in 1979/80 returned a total of 80 ±10 for the whole country, which effectively means along the Liberian border with a few in the other localities mentioned above. He considers that there could be as many as 50 pygmy hippos in the Loma Mountains area but his local informants may have been confusing their species.

Conservation status: The pygmy hippo is thought to be persecuted by local people because of the damage it causes to riverside vegetable gardens. Enforcement of protection laws is poor.

THE DISTRIBUTION AND NUMBERS OF *HIPPOPOTAMUS AMPHIBIUS* IN WEST AFRICA

Benin

Distribution: Sayer & Green (1984) mention the hippopotamus in their survey of large mammals in Benin but as their data refer to conditions in the period 1974–79, the situation may well have changed by now. They provide a map which shows hippos throughout the length of the Ouémé, Alibori, Mékrou and Pendjari Rivers. Hippos are also found on the floodplain of the Pendjari River, which extends into the neighbouring Burkina Faso. Hippos show seasonal movements between temporary wallows and rivers. These movements can be quite extensive for two hippos moved 30 km across country from a pond to a river (A.A. Green, *in litt.*). Hippos tend to move upstream on small tributaries in the rainy season.

Numbers: Hippos are nowhere abundant but 488 were counted in January 1979 along the Pendjari River and its lagoons downstream from its exit from the Atacora Mountains. Green (1997) put the dry season total along the Pendjari River on the border with Burkina Faso as being over 500 during the 1970s with concentrations of up to 70 in the lagoons. In March 1979 some 99 hippos died in four of these lagoons from unknown causes, but neither anthrax nor starvation was thought to be responsible. This represented a loss of around 40%. Later counts made on the Pendjari river in 1987 revealed only 441 hippos, mostly within Benin (J.A. Walsh, *in litt.*). Verschuren *et al.* (1989) reported a total of only 280 in the 1980s so it seems that the population did not recover. Elsewhere, such counts as have been made suggest that numbers are falling drastically, e.g. 12 on Ouémé River in 1977 but only four in 1988. It is now probably extinct on the Mékrou River. One was seen on the Alibori River in 1988 but only spoor were found a year later. Thirty-one were recorded on the Sota River in 1979 but there has been a sharp decrease since. A group of 53, seen in the Mona River on the Togo border in 1986, appears to be stable. It is unlikely that the total number in the country exceeds one or two hundred but as hippos move freely between Benin and neighbouring countries, a precise figure is meaningless.

Conservation status: The species seems to be decreasing everywhere throughout this small country. It is mainly confined to protected areas of which the most important centre is the Pendjari lakes system, which is included in the Boucle de la Pendjari National Park. The hippo is also said to occur in Benin's other national park, Parc National du "W", which is shared with Niger and Burkina Faso. The level of law enforcement appears to be poor and the outlook for the hippo in Benin is not hopeful. There is some hunting for meat, particularly along the Pendjari and Sota Rivers.

Burkina Faso

Distribution: The hippo is found in several national parks and reserves including the international "W" National Park and the Reserve Total de l'Arly. It also occurs on the river systems of Volta Noire/Grand Balé and Comoe/Leraba as well as in the appropriately named lake Mare aux Hippopotames.

Numbers: Information on numbers is scrappy but 80 hippos were counted within 1000 km² in and around the Arly reserve in 1973/74 (Green, 1979) and 221 in 100 km of the Pendjari/Mékrou river systems in 1981 (Bousquet & Szaniawski, 1981). C.A. Spinage (*in litt.*) puts the 1981 total in this part of Burkina Faso as 280. He also puts the 1981 totals in the Comoe/Leraba river system as 68 and in the Mare aux Hippopotames as 45. Adding these estimates together gives a total for the whole country of 403 although this does not include those in the "W" National Park, which move into Benin or Niger. There were also a few on the Volta Noire/Grand Balé river systems but no precise figure is known. Only three were reported there in 1981 with a further 16 in the nearby River Serou but 70 were present in 1975 (J.F. Walsh, *in litt.*). A round figure of 500 for the whole country would not be excessive but this applies to the situation in the early 1980s and not necessarily to present-day conditions.

Conservation status: There is cause for concern over the future of the hippo in Burkina Faso as numbers are low and are declining everywhere. There are several national parks and reserves that contain, or contained, hippos and most hippos are confined to national parks but law enforcement is weak. The most important region is around the Comoe River, which forms the border with the Ivory Coast. Numbers there were said to be stable in 1989. The species has enjoyed full protection throughout the country since December 1980 but that has not stopped it being shot for food.

Cameroon

Distribution: Not much is known about the distribution of hippos in Cameroon. They occur in and around Korup National Park in the south-west of the country where a survey by H. Cooper and R. Jensen (*in litt.*) recorded some on the Bake River near its confluence with the Miri River and a little further upstream near Bakut village. Other hippos occur in four national parks in the north, namely, Benoue, Bouba Ndjida, Faro and Kalamaloue, and in Lake Maga on the floodplain of the Logone River. As much of Cameroon is heavily forested, there are few suitable feeding areas for hippos.

Numbers: There is very little information on the numbers of hippos in Cameroon. Those in the Korup National Park probably do not exceed a few dozen and little is known of the numbers in the north of the country but P. Elkan (*in litt.*) mentions that there are at least 40 in the immediate vicinity of Bourmi on Lake Maga. He considers that the population in the lake is "substantial" but it is unlikely that there are more than a couple of hundred, which is probably close to the total for the whole of Cameroon.

Conservation status: Nothing is known about the conservation of the hippo in Cameroon. It is protected in the national parks but there is no information about the effectiveness of such protection. P. Elkin (*in litt.*) mentions a degree of animosity towards hippos because of their crop-raiding habits.

Central African Republic

Distribution: The hippo is widely distributed throughout the country except in the extreme north. Its distribution was probably once continuous but the population is now broken into isolated units.

Numbers: No country-wide survey has ever been made but a few counts are available from certain areas which suggest that numbers are everywhere decreasing. There were formerly some 1500 in Monovo-Gounda-Saint Floris National Park, including 700–1000 on Lake Gata, but no more than 100 are left (Barber *et al.*, 1980; A.A. Green, *in litt.*). A group of 136 was counted in Bamingui-Bangoran National Park in 1976 (Spinage *et al.*, 1977) but only 20 were recorded there in 1988 (A.A. Green, *in litt.*). The total for the whole country was probably several thousand in the 1970s but it is unlikely that more than a few hundred remain. J.M. Fay (*in litt.*) considers that the population declined by 75% in the six years from 1983 to 1989.

Conservation status: Although the hippo is fully protected by law, the level of enforcement is negligible except at Ozanga-Ndoki and Manovo-Gounda-Saint Floris National Parks, where there are overseas projects in operation. The presence of foreigners in the parks seems to act as a deterrent to poaching, which is otherwise uncontrolled. As well as hunting for meat, there is also a trade in teeth and hides, in which Sudanese intruders as well as local people are involved. The Sudanese use the hide to make bridles for their horses (J.M. Fay, *in litt.*).

Congo (Brazzaville)

Distribution: I. Nganga (*in litt.*) says that the species is localised but that where it occurs, it is widespread and numerous on suitable rivers. IUCN/UNEP (1987) lists Odzala National Park, Nyanga North Reserve and the Louna and Lesio Rivers in the Lefini Reserve as containing hippos. Spinage (1980), on the other hand, mentions their occurrence only on the Nyanga River.

Numbers: There is no published information on the number of hippos in the Congo. As the country was originally almost entirely forested, it is unlikely that the population was ever large and the few that remain are probably no more than a few score in number.

Conservation status: Nganga considers the conservation of the hippo to be satisfactory but Spinage says that there is little effective protection of wildlife in the country. The Nyanga North Reserve was established mainly for the protection of the species but it consists only of two narrow strips on either side of the Nyanga River.

Equatorial Guinea

Distribution: Not much is known about the wildlife of this small country but according to J.J. Ballesta and J.C. Bolibar (*in litt.*), hippos occur on the Campo River from near its mouth to the Yengue's Falls. There are now no hippos on the large Bioko Island, which forms part of Equatorial Guinea, and there is some doubt that they ever were present (T. Butynski, *in litt.*).

Numbers: No counts are available but the species is rare and thought to be decreasing. The total population is unlikely to reach three figures.

Conservation status: The status of the hippo is said to be "probably satisfactory" but there appear to be no conservation measures in operation. Some wildlife laws were, however, passed in 1989 giving the species full protection. There is disturbance from timber and fishing activities, which does not auger well for the long-term conservation of hippos but at least the local people do not hunt them.

Gabon

Distribution: Hippos are not found throughout the country because most of it is still heavily forested, but they occur along much of the coast and up the Ogooué River for a considerable distance inland – at least as far as Lopé Reserve. It is also said to occur in Sette-Cama and Moukalaba Reserves as well as in Wanga-Wongue National Park (M. Nicolle, *in litt.*).

Numbers: No information is available on numbers although the hippo appears to be locally common where it does occur. It is unlikely, however, that the total population is more than a few hundred.

Conservation status: The conservation status of the hippo in Gabon is not clear but informants in the country suggest that it is satisfactory, usually with a couple of question marks added. Numbers may be stable or could be decreasing slightly. The species is protected throughout the country and most of the animals occur within protected areas. Some are poached for meat and a few are killed by fishermen in self-defence (A. Blom, *in litt.*).

The Gambia

Distribution: The Gambia consists of little more than a strip of land on either side of the river. The hippo is confined to the freshwater stretch of the River Gambia adjacent to rice fields in the eastern half of the country. Occasionally, some hippos may wander farther downstream to saline water near the river's mouth (A. Camara, *in litt.*).

Numbers: Some counts were made by K. Pack (*in litt.*) in 1987/88, with a minimum count of 19 and a maximum estimate of 40.

Conservation status: With such a small population, the hippos are at risk from minor perturbations in their environment. There is antagonism towards them from rice farmers, whose crops are sometimes trampled or eaten by the animals. Numbers seem to be decreasing as fewer sightings have been reported recently. The hippo is fully protected and some occur on the tiny Baboon Island National Park and the regulations are enforced as far as resources and personnel permit. Hippos are treated with respect by the local people and are not traditionally hunted, except by older experienced hunters. Despite this, the future of the species in the Gambia must give cause for concern.

Ghana

Distribution: The hippo used to occur in most of the rivers in the north of the country but now it is probably limited to the Mole, Bui and Digya National Parks with some remnants along the Black Volta and Kulpawn Rivers (B.Y. Ofori-Frimpong, *in litt.*).

Numbers: The only information on numbers comes from J.F. Walsh (*in litt.*) who counted 32 on the Black Volta in February 1989. Assuming that other areas where hippo occur were of similar density, a total of a few hundred is the most that can be estimated for the country-wide total.

Conservation status: The species is fully protected in Ghana and the situation is said to be satisfactory by Ofori-Frimpong although he does mention some poaching for meat.

Guinea

Distribution: The hippo occurs on many rivers and streams throughout the country (L.Y. Macky, *in litt.*).

Numbers: No figure is available for the total population but it is quite numerous on rivers according to Macky. As an example, a minimum of 69 was counted in May 1989 on the River Sassandra, which forms the border with the Ivory Coast (J.F. Walsh, *in litt.*). Possibly, there could be one or two thousand in the whole of the country.

Conservation status: This is said to be satisfactory by Macky. The species is fully protected throughout the country but no indication was given of the effectiveness of the conservation laws. There were no reports of commercial exploitation and the principal threat to survival is destruction of habitat.

Guinea Bissau

Distribution: Guinea Bissau is a small country but it is well-watered and hippos are widely distributed (P. Chardonnet, *in litt.*). They occur particularly on the following rivers: Buba, Cacheu, Cacine, Corubal, Geba and Mansoa. They are also widely distributed on the off-shore Archipelago des Bijagos, where some are found in salt water. Islands with good densities of hippos include Bubaque, Orango and Orangozinho but Canhabaque, Caravela, Formaosa, Meneque, Poilao, Uada, Uno, Uracane and Unhocomo also contain a few hippos.

Numbers: No information was obtained on the number of hippos in the country but they were said to be locally abundant in the early 1980s (P. Chardonnet, *in litt.*). The population on the Archipelago of Bijagos was put at between 135 and 270 in an unpublished IUCN report. The total for the whole country might well run into four figures.

Conservation status: Chardonnet considers that the population was declining in the early 1980s and he expresses concern over illegal exploitation for meat with the sale of tusks as a sideline. Hippos are hunted by fishermen on the archipelago as well as by poachers on the mainland. The meat is not eaten by Muslims, presumably because of the relationship with pigs. It is surprising to find such taxonomic precision in a religious practice. At the time of the survey there were no national parks in Guinea Bissau but the hippo is officially protected everywhere. There is no information about the effectiveness of the protection.

Ivory Coast (Côte D'Ivoire)

Distribution: The distribution of the hippo in the Ivory Coast is rather biased towards the north of the country but hippos are found on several rivers including White Bandama, Bandama, Bou, Comoé (on the Burkana Faso border), Marahoué, Nzi and Sassandra (on the Guinea border). They also occur in coastal lagoons.

Numbers: A few counts were made in 1989 by J.F. Walsh (*in litt.*). The 69 seen on the Sassandra River have already been noted under the Guinea report. Counts on other rivers were: Bou 50, Comoé 720, Marahoué 18, Nzi 32 and White Bandama 46. The total on the Comoé River is much higher than the 68 reported for the same river in Burkina Faso. Some earlier counts in the Ivory Coast returned figures of 113 on White Bandama in December 1975 and seven on Bandama in July 1987. It is likely that the total in the whole country exceeds 1000 animals.

Conservation status: The hippo is protected within the national parks but not elsewhere. The outlook for the species in the country is thought to be poor. The White Bandama River system is the most important area but hippos are not protected there and numbers are declining through poaching. The greatest threat to the species appears to be the expanding human population.

Liberia

Distribution: The common hippo has been reported as having occurred in the recent past in the Lofa River but the records are of doubtful validity (M.E.J. Gore, *in litt.*). If it ever did occur in Liberia, it almost certainly is not present now.

Mali

Distribution: The hippo occurs on a number of rivers in the south-west of the country. The regions to the north are too arid to support the species. It occurs on the Niger River near Bamako as well as on the following rivers: Bagoé, Bani, Banifing IV and Baoulé (J.F. Walsh, *in litt.*). It also occurs in Lake Fishpool some 25 to 45 km south of Gao.

Numbers: A few scattered counts are mentioned by Walsh. These include 10 on R. Bagoé in July 1975 and 50+ on R. Baoulé in June 1979. Hippos were recorded as "present" near Bamako in 1988 but the only evidence of their existence in Banifing IV was spoor. "A few" were reported in R. Bani in June 1970 by J. Henderickx (*in litt*). There are two records from Lake Fishpool with 1 hippo seen in June 1978 and 2 in October 1978. There are clearly not many hippos in Mali and a total of 100 might not be far wrong.

Conservation status: There is no information on conservation measures in the country. The hippo occurs in the Boucle du Baoulé National Park so it theoretically enjoys some protection but how far the laws are enforced is unknown. The main threat to the species seems to be the general desertification of Mali.

Niger

Distribution: Hippos occur on the River Niger with 80% of them between the villages of Ayerou and Firgoun in the south-west of the country (J.E. Newby, *in litt.*). It was formerly present in Lake Chad but disappeared when the waters receded between 1974 and 1987. There was some expansion of the water in 1989 but no reports of hippos returning had been received at the time of this survey.

Numbers: Newby gives a total of about 200 for Niger, mainly on the river, but J.F. Walsh (*in litt.*) found very few on the river with counts of only six in 1976 and of one in 1988. These were probably not extensive searches and Newby's figure of 200 is accepted.

Conservation status: The population is thought to be decreasing and giving cause for concern. The official protection given to the species is fair but political pressure to deal with crop-raiding animals reduces the effectiveness of the protection. There is some hunting but falling water levels in the Niger River and habitat destruction are thought to be the principal threats to the hippo's survival.

Nigeria

Distribution: The hippo is widely distributed throughout Nigeria, occurring in Kainji Lake National Park and a number of game reserves including Gashaka-Gumti, Hadejia Wetlands, Kwiambana and Yankari. It is also found in the Benue River.

Numbers: There seem to be only some 200 hippos in Nigeria with no more than 100 in Yankari Game Reserve and up to 56 in Kainji Lake National Park (F.O. Amubode, *in litt.*; A.A. Green, *in litt.;* Marshall, 1985; Sikes, 1974). Comparable numbers were counted in other game reserves, e.g. <30 in Gashaka-Gumpti and 24 in Kwiambana.

Conservation status: One informant said that the situation was satisfactory and not giving cause for concern and the other said the exact opposite. Hippos are probably increasing in Lake Kainji National Park but not elsewhere. The low numbers should give cause for concern. The hippos are well protected in the national park but elsewhere are persecuted for food and as agricultural pests. The main threats to survival are meat poaching and loss of habitat.

Senegal

Distribution: The hippo was once found throughout Senegal but it is now confined to the east and south of the country (A.R. Dupuy, *in litt.*). The bulk of the population resides in the Niokolo-Koba National Park on the upper reaches of the Gambia River and its tributaries. A few that live in the estuary of the river in Basse-Casamance National Park spend the day in salt water.

Numbers: K. Pack (*in litt.*), in a review of the literature, deduced a total of 800 for the Niokolo-Koba National Park but Dupuy gives a more conservative figure of 500 for the whole country, which is virtually the same as for the park as far as hippo numbers are concerned. In an earlier account, Dupuy (1971) put the park's total at 700 and mentioned an earlier estimate of 200 made in 1957. A best estimate for the present population might be in the order of 500 although it could be a few hundred more.

Conservation status: According to A.R. Dupuy (*in litt.*), the population has been falling at a rate of 6.5% per annum over the ten years prior to 1989 and the future of the species in Senegal is giving cause for concern. It is theoretically protected but there is little in the way of law enforcement, except, perhaps, in the Niokolo-Koba National Park. The main threat to survival is hunting for the "ivory" trade.

Sierra Leone

Distribution: The hippo is mainly confined to protected areas in the north and east of the country with the majority on the Greater and Lesser Scarcies Rivers in the Outamba-Kilimi National Park.

Numbers: A survey made in 1979/80 suggested a country-wide total of 160 ± 30 (G. Teleki, *in litt.*). Most of these (*c.* 60) were in the Outamba-Kilimi National Park. G. Davies (*in litt.*) mentions old reports of hippos in Lake Kase (1932), Rokel River (1960) and on the coast at Sulima (1932). Numbers have certainly declined recently and a present-day total of 100 is probably optimistic.

Conservation status: The future of hippos in Sierra Leone is giving cause for great concern. Numbers are declining rapidly and the species is at risk of extinction. Although it is fully protected, the laws are rarely enforced. Hippos are killed for their meat and tusks and in defence of crops.

Togo

Distribution: This small country does not have much space for many hippos but the species occurs on the Moto and Oti Rivers (J.F. Walsh, *in litt.*). It is also found in the Keran National Park (IUCN/UNEP, 1987).

Numbers: The results of a few counts are available from Walsh. Fifty-three were reported in 1986 on the Mono River on the Benin border, as was mentioned in the account for that country. Some of the 441 counted in 1979 on the Pendjari River system in Benin were presumably in Togo on the Oti River. Elsewhere a few exist on the Koumongou River in the Keran National Park. It is difficult to give a figure for the total population in Togo because many of the hippos live in rivers that are international borders, but there can be only a few score within the country itself.

Conservation status: The species is fully protected and enforcement of regulations seems to be rather better than in many West African countries. Numbers are thought to be stable on the Mono River.

SUMMARY OF THE NUMBERS OF THE COMMON HIPPO IN WEST AFRICA

The distribution of the common hippo is shown in Fig. 10.3 and Fig. 10.4 for west and west-central Africa respectively. Few of the informants were prepared to put figures to the number of hippos in the various countries but some tentative estimates, based on the above analyses are included in Table 10.1. Simple totals should probably

Figure 10.3   Distribution of the common hippopotamus in western West Africa in 1989 as determined from the IUCN SSC investigation.

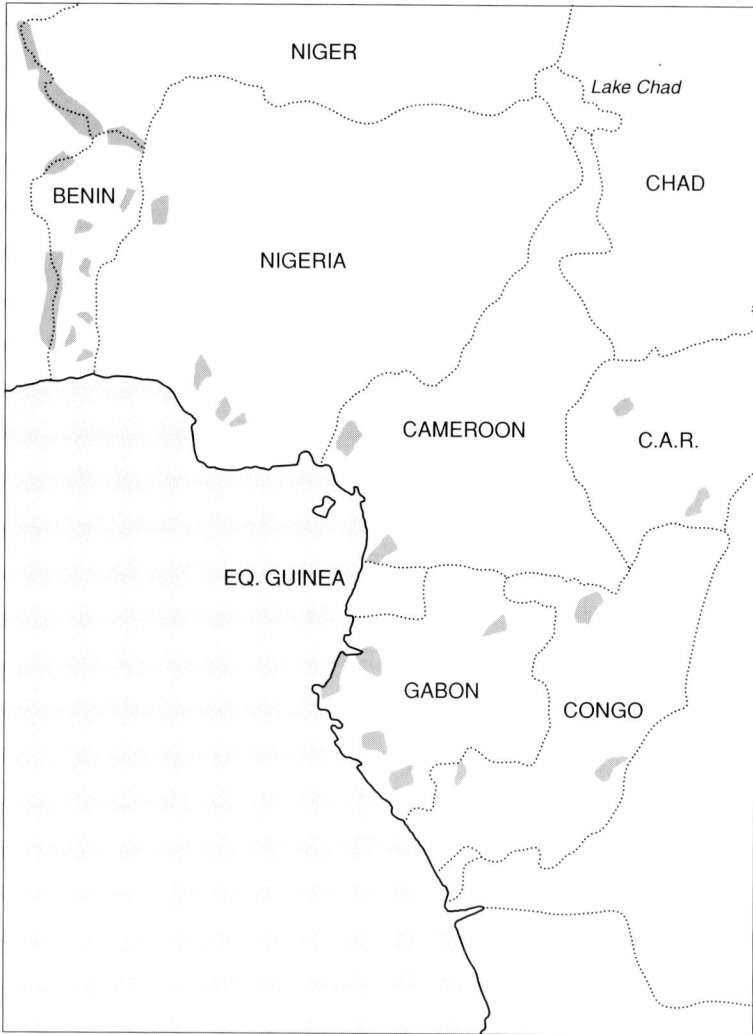

Figure 10.4   Distribution of the common hippopotamus in central West Africa in 1989 as determined from the IUCN/SSC investigation.

not be given, since they are sure to be incorrect, but as long as the figures are treated with due caution, they do provide a means of assessing the overall status of the species in Africa.

West Africa would not be expected to contain many hippos as the original vegetation over much of its area was rain forest, which is not prime habitat for the species, and although some of the forest has been converted to savanna through human activities, there was no existing reservoir of hippos that could have colonised the new grass-

Table 10.1   Estimates of hippo numbers in West Africa.

Country	Numbers	Conservation Status	Population Trend	Protection/ Enforcement	Concern?
Western Group	**3640**				
Gambia	40	RD-LD	D?	G/F	Y
Guinea	2000	W-LA	D	G/G	Y
Guinea Bisau	1000	RD-LA	D?	G/U	Y
Sierra Leone	100	RD-LD	D	G/P	Y
Senegal	500	RD-LA	D	G/P	Y
Central Group	**2100**				
Benin	150	RD-LA	D	U/P	N
Burkina Faso	500	RD-LD	D	H/P	Y
Ghana	300	RD-L	D?	G/G	Y
Ivory Coast	1000	RD-LD	D	H/P	Y
Mali	100	RD-LD	U	U/U	Y
Togo	50	RD-LD	S?	G/G	N
Eastern Group	**1450**				
Cameroon	200	W-LD	U	G/F	Y
C.A.R.	850	RD-LA	D	G/P	Y
Niger	200	RD-LA	D	G/F	Y
Nigeria	200	RD-LD	D	H/F	Y
Southern Group	**400**				
Congo	50	W-LA	I	G/U	N
Equatorial Guinea	100	RD-LD	U	G/P	N
Gabon	250	W-LD	D?	G/P	N

Total for West Africa – 7590

KEY

Conservation Status: LA = locally abundant; LD = low density; HD = high density; RD = restricted distribution; W = widespread.

Population Trend: D = decreasing; I = increasing; S = stable; U = unknown.

Protection: G = total protection; H = partial protection; U = unknown.

Enforcement: E = excellent; F = fair; G = good; P = poor.

Concern? (over population status): Y = yes; N = no.

lands. Those countries in the far west, particularly Guinea, Guinea Bissau and Senegal, contain the bulk of the population with over 3600. The next most populous group of states is the central group with over 2000 altogether, half of which are in the Ivory Coast. Each of the countries in the eastern group contain a few hundred hippos with most in the Central African Republic. Fewest of all are in the three southern states of Congo, Equatorial Guinea and Gabon. It will be noticed that the population sizes decline from west to east and from east to south. These compass directions are, of course, relative, as south in this context is equatorial. Liberia is not included in the list as the hippo does not occur there and no hippos were reported from Mauritania, where the species once occurred.

Adding together the totals for individual countries produces a sum of 7590 for the whole of West Africa. No information was received from Chad during my survey but the population was reported to be 400 at the 1994 CITES meeting when the species was placed on Appendix II (M. Haywood, pers. com). This brings the total for West Africa to nearly 8,000. As mentioned above, a precise figure for the number of hippos is bound to be incorrect but it does serve as an indication of the probable size of the population. At best one can say that there are several thousand hippos in the region but not tens of thousands.

SUMMARY OF THE CONSERVATION STATUS OF THE COMMON HIPPO IN WEST AFRICA

Of the 18 countries with hippos, numbers were thought to be declining, or probably declining, in 13, probably stable in one and increasing in another. The situation in the three remaining countries is unknown. These, like the other conservation criteria, represent opinions and may not necessarily be correct.

Concern over the future of the hippo was expressed in most states except for the three in the southern group, which may reflect optimism in the informants rather than the actual situation. These, and the other two countries where no concern was felt, contain the fewest numbers and wherever there were reasonably sized populations, the long-term survival was in doubt. The degree of legal protection was good or fairly good in all countries for which an opinion was given but the enforcement of this protection left much to be desired. Only in three countries was it considered good, in four it was thought to be fair but in eight law enforcement was considered to be poor.

The distribution of the population was said to be restricted in 14 of the countries and the density low in 11 of them. The hippos were believed to be widespread in four countries but only in Guinea did this refer to a substantial number of animals. There were no high density regions but the species was considered to be locally abundant in seven countries. A problem here is the interpretation given to "abundant" whose meaning can vary considerably from one observer to another.

The overall conservation status of the hippo in West Africa is poor. Total numbers are low and such as there are are split into small, isolated groups of a few hundred at most. Populations as small as these are vulnerable to catastrophes, such as drought or a surge in poaching, as well as to genetic problems resulting from the risk of inbreeding so that the prospects for their long-term survival must give cause for concern. As some of the assumed subspecies are confined to West Africa, there is a risk of their extinction.

CHAPTER 11

THE DISTRIBUTION AND NUMBERS OF *HIPPOPOTAMUS AMPHIBIUS* IN EAST AFRICA

Hippo under water at Mzima Springs

INTRODUCTION

The same procedure will be adopted as for West Africa, i.e. each country will be considered in turn. Zaire, which changed its name in 1997 to the Democratic Republic of the Congo, is included here although it is as much a west as an east African country. Its hippos, however, are mainly in the eastern regions of the country so it makes sense to treat the country as being in East Africa.

Burundi

Distribution: The hippo is said to be restricted in distribution but to be abundant where it does occur (P. Chardonnet, K.M. Doyle, P.C. Trenchard, *in litt.*). Its principal haunts in Burundi are the Ruvubu River in the national park of that name, Rusizi Nature Reserve and the Malagarazi River, which is on the Tanzanian border in the south-east of the country. It also occurs on a few sites along the shore of Lake Tanganyika.

Numbers: Estimates made by Chardonnet suggest that there are probably several hundred in the country altogether, e.g. a "few hundred" on the Ruvubu River, 200+ on the Ruzizi River and "a good population" on the Malagarazi River. At a conservative estimate, Trenchard puts the country's total at 1000+. Doyle made a count in 1990 on a 120 km stretch of the Ruvubu River, of which 98 km were within the national park and counted only 39 hippos, all but two in the park. Previous estimates for the park have varied between 500 and 1500 and although the many wallows close to the river were not surveyed, it does seem that the population has declined recently. With some trepidation, therefore, the figure of 1000 is accepted for Burundi's hippos.

Conservation status: Chardonnet believes that the population is slowly decreasing but Trenchard thinks it is increasing, even outside protected areas, thanks to the enforcement of hunting laws. The regulations do seem to be applied and no hunting is allowed. Nevertheless, hippo teeth are on sale in the capital Bujumbura although it is not known whether they originate in Burundi. The main threat to survival is loss of habitat and the hostility from farmers whose crops are destroyed by hippos. All these remarks apply to the situation in the country before the political problems of the mid-1990s and it is anyone's guess what the fate of the hippos has been since then.

Ethiopia

Distribution: The hippo is widespread in regions up to 2000 m in elevation. Its main strongholds seem to be the Omo, Awash and Great Abbai (Blue Nile) Rivers. It also occurs in most of the larger lakes and, as isolated populations, in smaller swamps and pools. Its northern limit is the Setit River. In the arid south-west, it is confined to Webi Shebeli and Setit River. This information is available in Yalden *et al.* (1984) whose map is reproduced in Fig. 11.1.

Numbers: No recent counts have been made but the hippo is said to be numerous. There are not many in the south-east. The total number probably runs into several tens of thousands and an educated guess of 5000 is more likely to be too low than too high.

Conservation status: There is very little information on the conservation of the hippo in Ethiopia. There is some hunting in certain areas but probably not enough to present a real threat to survival. Numbers may have declined in places but in general they appear to be stable.

Kenya

Distribution: The hippo is widely distributed throughout Kenya and is found on most major waters, particularly in the south of the country. It occurs in all the big lakes including Baringo, Naivasha, Nakuru, Turkana and Victoria. It also occurs on the following rivers; Athi, Galana, Mara Nzoia, Tana and the Uaso Njiro River upstream of

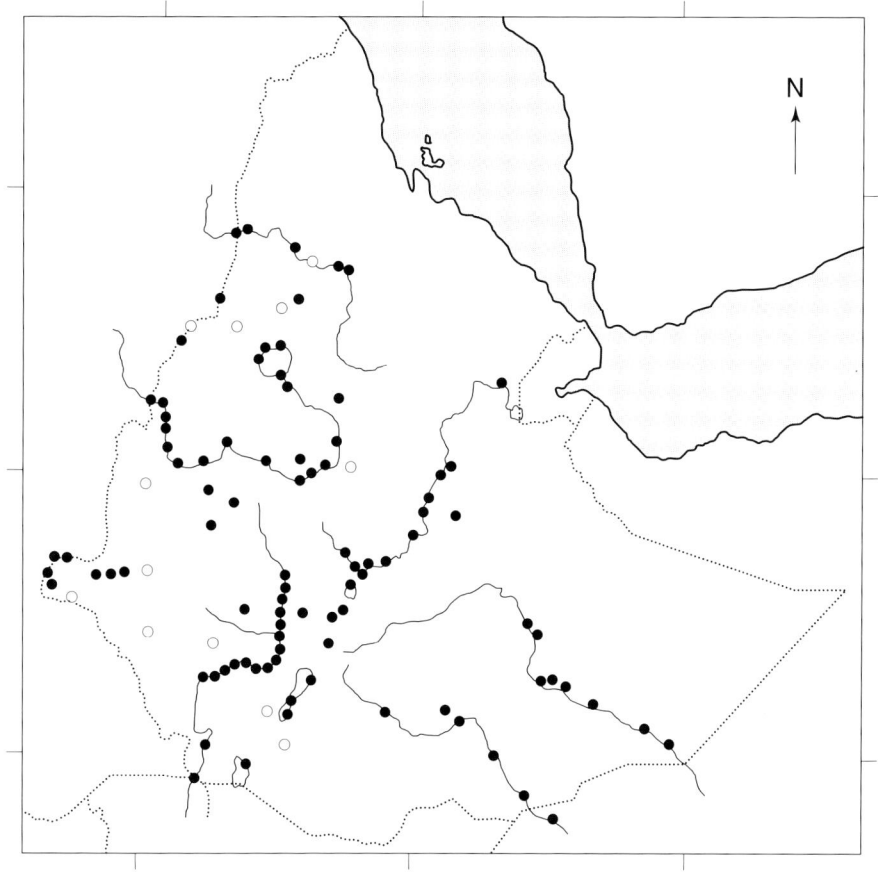

Figure 11.1 The distribution of the common hippopotamus in Ethiopia. From Yalden *et al.* (1984). ○ = locality of specimen approximate or doubtful; ● = locality of specimen precisely identified. Both types refer to the definite presence of hippos.

Lorian Swamp. Other bodies of water, such as the Mzima Springs and the Amboseli swamps, hold hippos and some coastal sites are known. Fig. 11.2 is a map showing the distribution in the early 1960s and it is likely that a more up-to-date map would be much the same.

Numbers: There have been no country-wide counts of hippos in Kenya but there are some detailed published counts on certain waters. A. Smart (*in litt.*) carried out a three-month survey in 1988 of Lakes Naivasha and Oloidien in the Rift Valley and counted a total of 228 hippos along a shore line of about 70 km. Coe & Collins (1986) report a total of 220 sightings of hippos from the air along the Tana River from Osako in the east to the Adamson Falls. Olivier & Laurie (1974a) calculated a total of 1189 hippos from corrected aerial counts made in 1971 in that part of the Mara River within the Serengeti and 738 in the Kenyan section of the

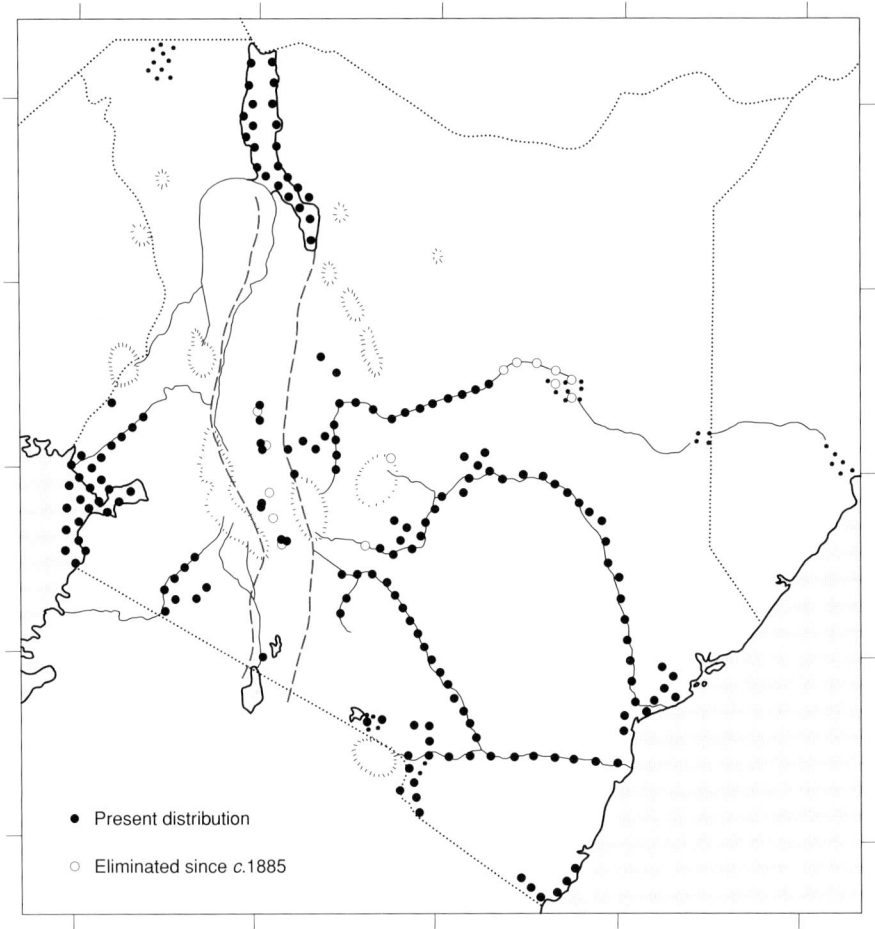

Figure 11.2   The distribution of the common hippopotamus in Kenya. From Stewart & Stewart (1963). ● = distribution in 1963; ○ = eliminated since c. 1885.

river, a total of 1927. This is close to the 2132 counted from the ground on the whole Mara River system. Karstad & Hudson (1984) made a series of aerial and ground counts in 1980 and 1982 on 124 km of the Mara River in Kenya between Emarti and the Tanzanian border. They recorded about 1800 hippos on this stretch but from ground/air comparisons they put the true total at 2819. If this is correct, it means that there has been an increase in numbers, which Karstad & Hudson consider, from historical evidence, was at an annual rate of 16.5% between 1959 and 1971 and of 10.3% between 1971 and 1980, dropping to 0.9% between 1980 and 1982. Although the hippo is widespread in Kenya, it does not occur anywhere in large numbers and a conservative total of 5000 is suggested.

Conservation status: The conservation of the hippo in Kenya seems to be satisfactory although there are no hard data to support this belief. It is generally well protected in national parks and reserves, where the majority of hippos reside, although the level of protection varies from place to place. The main threat to the species is the extension of agriculture, which is reducing the amount of grazing land available.

Rwanda

Distribution: The hippo was once a common species on the Akagera River, which forms the border with Tanzania, and in the adjacent lakes and swamps of the Akagera National Park. The civil wars of the mid-1990s may have affected this distribution but no up-to-date information is available.

Numbers: No recent surveys have been made but in 1969 Spinage *et al.* (1972) counted 671 hippos in the park and a further 80 in the adjoining Mutara Game Reserve. There were probably some more outside these protected areas but if so, they were unlikely to be more than a few. The total for the whole of Rwanda is now probably less than it was in 1969 and a figure of 500 is suggested.

Conservation status: No information was obtained from the 1989 questionnaire but the prospects for the survival of hippos in Rwanda are not good. The recent guerrilla activity and the known infiltration of people into the national park have undoubtedly affected the hippos. Those outside the park are few and widely scattered and may not exist for much longer in this densely populated country.

Somalia

Distribution: Hippos are confined to the two rivers of Somalia – Juba and Scebele (Fagotto, 1985).

Numbers: The population of hippos in Somalia is very small and the species is close to extinction in the country. Groups of from four to six were reported from the lower reaches of the Scebeli by Fagotto (1985), who mentions that the species was more numerous along the Juba River. The country-wide total at the time was probably measured in dozens.

Conservation status: The hippo's status in Somalia is very poor and the outlook is even worse. Fagotto says that it was being slaughtered in the 1980s by poachers for its tusks and by farmers in defence of their cultivations. It is likely that by now very few remain, perhaps 50 at the most.

Sudan

Distribution: The hippo is widespread in southern Sudan and occurs in most rivers south of Malakal, particularly the Nile, Obat and Jur. On the Nile itself the hippo extends as far north as Kosti. It also occurs in the Sudd and in tributaries of the Nile.

Numbers: No recent counts have been made in the Sudan and none is likely in the near future owing to the political disturbances in the region. M.B. Namir (*in litt.*) considers that the civil war has resulted in an increase in the hippo population because the people have left as refugees for safer regions in the north of the country. There is no doubt that the species was once very common in southern Sudan and it probably still is. It is impossible to give a precise figure but hippos can probably still be numbered in thousands. For the purposes of deriving a total for the region, an estimate of 6000 is suggested.

Conservation status: Despite the wars and upheavals in the region, the outlook for the hippo is by no means entirely gloomy. It is undoubtedly being hunted for meat and this is the most serious threat to its survival. In theory, at least, it is well protected as many of its haunts are national parks or reserves (Hillman, 1982) but there is no information on how well these areas are being policed. National parks containing hippos include Southern, Nimule and Boma. Game reserves with hippos are Mongalla, Badingeru, Zeraf, Fanyikang, Juba and Shambe.

Tanzania

Distribution: The hippo is a common species throughout Tanzania with the largest numbers being found in the Selous Game Reserve. Most national parks and reserves with suitable habitat contain the species.

Numbers: Detailed counts have been made in the Selous Game Reserve. An aerial survey along the Rufigi River in 1990 counted 6866 hippos, a figure which, when extrapolated, resulted in a total of 20 598 for the whole reserve (Games, 1990). Numbers recorded on two earlier aerial counts using a different technique involving straight-line transects produced similar results with 15 843 in 1986 and 24 169 in 1989. I. Douglas-Hamilton (*in litt.*) made an aerial survey over 74 000 km^2 of the Selous in 1986 and returned a total of 16 900 with a standard error of 6397. Presumably this was different from the 1986 count reported by Games. Some more restricted counts on limited stretches of the river suggest that there has been a decline in the upper reaches over the past 30 years. A survey in 1963 counted 20.8 hippos per kilometre between the Shiguli Falls and Stiegler's Gorge but by 1988 the density in this section of the river had fallen to 7.8 and by 1990 to 7.7. A. Samuels (*in litt.*) recorded 1894 hippos in July 1987 on 115 km of the Rufigi River between the Shuguri Falls and Steigler's Gorge. This is a linear density of 16.5 hippos per kilometre, not far short of the 1963 figure.

If a round figure of 20 000 is allowed for the Selous and 5000 for the rest of the country, a total of 25 000 is derived. This is something of a guess as there are no reliable counts over much of the hippo's distribution in Tanzania.

Conservation status: The species has full protection in the national parks although the degree of effectiveness of the protection is probably variable. The main threat to survival is hunting for meat and some poaching is known to occur in the Selous (A. Samuels, *in litt.*). On balance and given the large size of the population, the status of hippos in Tanzania may be considered to be satisfactory.

Uganda

Distribution: The Ugandan hippos occur mainly in the south of the country. They are common along the River Nile and its tributaries, the Kazinga Channel in the Queen Elizabeth National Park, the Semliki River and in the larger lakes, such as Victoria, Albert, Edward, Kyoga and George. Some hippos do occur in the far north, e.g. in the Kidepo Valley National Park, which borders the Sudan.

Numbers: The largest numbers of hippos used to occur in the vicinity of Lakes Edward and George in the Queen Elizabeth National Park but the populations were severely depleted in the 1980s through poaching. Nevertheless, a substantial proportion remains. This was an area where the hippo was considered to be over-abundant, resulting in the culling of 7000 in the 1960s as described on pp. 89*ff*. The park's population, previously 21 000, was reduced to 14 000 and remained at around that level for the next 15 years or so. An aerial sample count made in 1989, after the cessation of the poaching, estimated 2172 (R. Olivier, *in litt.*). The only subsequent counts appear to be the aerial surveys in 1995 by F. Michelmore who reported a total of 2800 in the Queen Elizabeth National Park and one of 1500 in Murchison Falls National Park. Schools of several hundred occur in Lake Victoria and other lakes in the south and a country-wide total of 7000 is probably not far out.

Conservation status: There is a significant local trade in meat, especially in Ankole, where the hippo is considered a delicacy. It is, however, fully protected in the national parks and the regulations are observed fairly well. Numbers are probably still decreasing but not drastically.

Zaire (Democratic Republic of the Congo)

Distribution: Much of Zaire is (or was) covered by rain forest, which is not good hippo habitat. Most of the hippos, therefore, live in the east where the country is more open. A few, however, are found in parts of the north-west. Some occur in the north-east in a few localities along the larger rivers of the Ituri Forest as well as on the Zaire River (Yangambi) and the Boma River (Carpaneto & Germi, 1989; J.A. Hart, *in litt.*). Their main stronghold, however, is around the Rift Valley lakes in the east, particularly in the Virunga National Park. Other national parks with hippos include Upemba, Salonga, Kundelungu and Garamba.

Numbers: There have been some good counts made in some of the national parks. By far the most occur in the Virunga National Park, where 22 875 were estimated from an aerial census made in 1989 by C. Mackie (*in litt.*). Mertens (1983) counted 21 095 in the park in 1981. A count in Garamba National Park, made in March 1988 by Mackie and Hillman-Smith, recorded 2851 (K. Hillman-Smith, *in litt.*). Numbers elsewhere have not been counted but the populations are probably small. A likely estimate for the whole country is around 30 000.

Conservation status: Until the upheavals caused by the troubles in Burundi and Rwanda in the mid-1990s, one would have said that the conservation situation was satisfactory. Numbers appeared to be stable in the national parks. With the influx of refugees into the Virunga National Park, the future of the hippos there must give rise to concern.

SUMMARY OF THE CONSERVATION STATUS OF HIPPOS IN EAST AFRICA

A summary of the estimated totals of hippos in the countries of East Africa is given in Table 11.1 and the distribution is shown in Fig. 11.3. The most important countries for hippos in terms of numbers are Zaire with 30 000 and Tanzania with 25 000. Together, these two countries account for nearly 70% of the region's estimated total of 79 550 animals. Substantial populations are also found in Ethiopia, Kenya and Sudan.

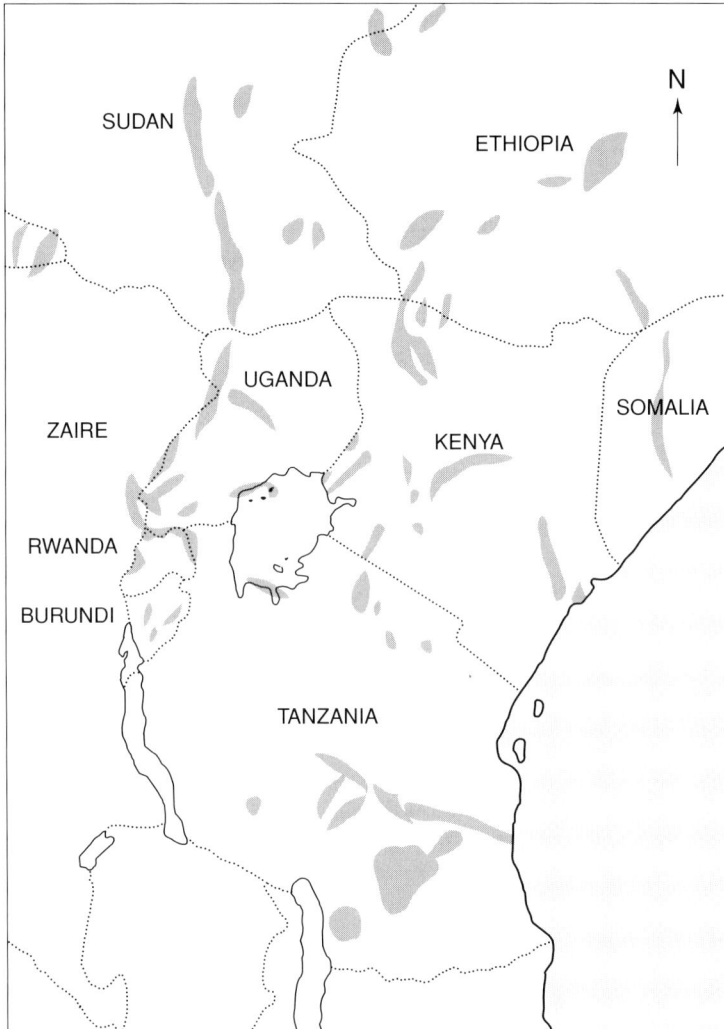

Figure 11.3   The distribution of the common hippopotamus in East Africa as determined from the IUCN/SSC investigation.

Table 11.1 Estimates of hippo numbers in East Africa.

Country	Numbers	Conservation Status	Population Trend	Protection/ Enforcement	Concern?
Burundi	1000	RD–LA	U	G/G	N
Ethiopia	5000	W–LA	S	G/F	N
Kenya	5000	W–LA	S	G/G	N
Rwanda	500	No recent information			
Somalia	50	RD–LD	D	U/P	Y
Sudan	6000	RD–LA	U	G/F	N
Tanzania	25 000	W–LA	S	G/F	N
Uganda	7000	W–LA	D	G/P	Y
Zaire	30 000	RD–HD	D	H/F	N

Total for East Africa – 79 550

KEY
Conservation Status: LA = locally abundant; LD = low density; HD = high density; RD = restricted distribution; W = widespread.
Population Trend: D = decreasing; I = increasing; S = stable; U = unknown.
Protection: G = total protection; H = partial protection; U = unknown.
Enforcement: E = excellent; F = fair; G = good; P = poor.
Concern? (over population status): Y = yes; N = no.

In the 1989 survey, concern over the hippo's survival was expressed in only two of the countries, Somalia and Uganda, and these were two of the three countries that were thought to have declining populations. The observers would probably revise their opinions today and include Sudan and Zaire in the list of countries with threatened hippo populations.

The distribution was restricted in four of the countries, including Zaire, which has the largest number. In another four countries, the species is widespread and abundant. Conservation legislation is adequate in most range states and enforcement of the law is fair to good except in Somalia and Uganda although the situation in Uganda has probably improved by now.

Despite the problems, East Africa is clearly an important region for hippos and there is no immediate risk of their extinction, although some populations may soon die out with a subsequent loss of genetic diversity. Tanzania is an important country for hippos with a large population that is stable, widely distributed and abundant, although law enforcement is said to be no more than fair. Recent military activity in Rwanda, Zaire and Sudan makes the situation very difficult to assess but it is seems inevitable that the hippos must have suffered to a greater or lesser extent.

CHAPTER 12

THE DISTRIBUTION AND NUMBERS OF *HIPPOPOTAMUS AMPHIBIUS* IN SOUTHERN AFRICA

Hippo wading in river

INTRODUCTION

This is the final sector of Africa to be examined and as before, each country will be considered in turn under the same headings.

Angola

This is the only important country for which no information could be obtained during the survey. This was due mainly to the unsettled political situation at the time, which precluded observers from moving around freely. It is known that the species was once widespread throughout Angola (Sidney, 1965) particularly in the east on the

Cunene, Cubango, Cuando, Cuanza, Longa and Zambezi rivers. It is unlikely that the hippo has disappeared from all these rivers, although there has probably been a significant reduction in numbers. For the purposes of amassing a total, a notional figure of 2000 is allowed for the present population. Although this may induce an error, it will probably be a smaller one than would be the case were the country to be ignored altogether.

Botswana

Distribution: Mainly confined to the north-west of the country, particularly in the Okavango Delta and in the Chobe/Linyanti river system. Outside this area, there is (or was) a small population near Ghanzi, which is still in the north close to the Namibian border (Smithers, 1971). Most of Botswana, essentially the Kalahari Desert, is too dry for hippos.

Numbers: C.A. Spinage (*in litt.*) put the total population in 1987 for northern Botswana at 1600 in the wet season and 500 in the dry season, with a few, some 16 or more, in the Limpopo River. A more recent aerial survey was made in the dry season of 1994 and returned an estimate of 2789 with 50% confidence limits, i.e. a range of 1395 to 4184 (J. Howes, *in litt.*). This survey was not stratified for hippo and was considered to be a significant undercount. Spinage's figure of 1600, therefore, is probably now too low and a total of 4000 is suggested as the population size for the country as a whole.

Conservation status: The hippo occurs in a number of protected areas in Botswana, including Chobe National Park and the Moremi and Makgadikgadi Pans Game Reserves. Numbers are probably decreasing rather than increasing but this does not seem to be giving rise to concern locally. The hippo is protected from hunting although some are shot by farmers if raiding crops. Enforcement of the law is said to be fair.

Malawi

Distribution: The hippo occurs in "all rivers and lakes of sufficient size" according to R.H.V. Bell (*in litt.*). The main concentrations are found on the Shire River, particularly in the upper reaches, and in Elephant Marsh on the lower stretches of the river near Mwabvi Game Reserve. Elsewhere, large numbers occur in Lake Malombe and the adjacent Liwonde National Park as well as in the south-west arm of Lake Malawi.

Numbers: A few aerial counts, mostly unpublished, have been made along the Shire River but no country-wide census has been undertaken. Bhima (1996) lists some of these and reports the results of an aerial survey that he made in October 1993 on a stretch of the Shire River from its outlet from Lake Malawi to Zalewa Bridge, a distance of some 291 km, including the 85 km shoreline of Lake Malombe. The total of 1234 was derived by applying a correction factor of 1.2 based on a comparison of aerial/boat counts made in Liwonde National Park. The density varied from 0.7 hippos km^{-1} of shoreline in Lake Malombe to 20.2 km^{-1} of river within Liwonde National Park. R.H.V. Bell (*in litt.*) puts the total for the whole of Malawi at 10 000 but warns that this is no more than an educated guess. In the absence of other information, however, this figure is accepted.

Conservation status: The conservation status of the hippo in Malawi appears to be satisfactory. All populations are thought to be stable but no serial counts have been made to allow trends to be assessed. The hippo is a fully protected species and may be killed only under licence, which is rarely issued except on control. The species occurs in several national parks and reserves although about 75% of the population live outside protected areas. The wildlife regulations are generally well enforced. The main threat to survival comes from the attitude of the local people, to whom the hippo is a pest, owing to its destruction of crops and the damage it causes to the nets of fishermen, who are often subjected to attacks. R.H.V. Bell (*in litt.*) considers that more people are killed by hippos in Malawi than by any other wild animal, with the possible exception of crocodiles.

Mozambique

Distribution: The only recent source of information on the hippos of Mozambique is that contained in a report by L. Tello (*in litt.*). Hippos occur in all the major rivers, particularly the Rovuma and Lugenda in the north, the Zambezi and Pungue in the centre and the Save River in the southern half of the country. They are also common in the rivers running into the sea near Maputo. The artificial lake on the Zambezi, created by the Cabora Basa Dam, supports a substantial population.

Numbers: Tello gives details of counts which show that the densest populations are to be found in Gorongosa National Park (3483 in 1972 and 3597 in 1977) and the Zambezi Delta (2000–3000). Other important regions include Tete Province (1500–2000) Niassa Province (1500–2500) and Cabo Delgado Province (1000–2000). There have been declines in some regions, attributable either to poaching or to drought, and present numbers in Niassa and Cabo Delgado Provinces, for instance, are nearer 500 and 1500 respectively. Nevertheless, there are still large numbers of hippos in Mozambique and Tello puts the total between 16 000 and 20 500. For the purposes of obtaining a regional total, a compromise figure of 18 000 will be taken.

Conservation status: Hippos have increased in the Zambezi Delta but elsewhere the picture is one of decline except in Tete Province, where numbers are stable. The degree of attrition is not known but the main cause of the decline appears to be illegal hunting exacerbated by drought. The military activity in the country over the past decades has hindered investigations but the national park structure has been under stress and the enforcement of the law is probably difficult. The situation gives cause for concern but the hippo population may be big enough to absorb the pressures and to recover when conditions improve. The most hopeful regions for the conservation of hippos are the Gorongosa National Park and the Zambezi Delta.

Namibia

Distribution: Hippos in Namibia are confined to the Caprivi Strip as the rest of the country is too dry to support them. They occur along that part of the Zambezi which forms the border with Zambia and stretches of the Linyanti and Kwando Rivers bordering Botswana.

Numbers: H.J.W. Grobler (*in litt.*) gives the results of some game censuses made between 1980 and 1990. The average number of hippos recorded was 238 with a maximum of 695 (in 1980) and a minimum of 24 in 1986. The 1990 total was 186. These fluctuations may reflect inaccuracies in counting but there was a marked decline until 1986, after which there appears to have been something of a recovery. A present-day total of about 400 for the whole country is suggested by Grobler.

Conservation status: The hippo is hunted for meat by the local people although it does have complete legal protection and the law is generally well enforced. It occurs within the Western Caprivi Game Reserve and the Nkasa Lupala National Park. It may also be present in the Mudumo National Park. Within the protected areas, it is generally safe but populations do occur outside them. The opinions of the correspondents are divided over whether there is cause for concern over the future of the species.

South Africa

Distribution: South African hippos are confined to the north-east of the country, mainly in the Transvaal and the northern tip of Natal. Most are in the Kruger National Park where they are found in perennial rivers as well as in dams and the larger pools of seasonal rivers. Some have been translocated from the Kruger to other parks and reserves further south. There are no hippos in the Orange Free State apart from six on game farms (S. Vrahimis, *in litt.*).

Numbers: Counts from the air were made annually in the Kruger National Park between 1984 and 1994 on the five main rivers in the park, the Sabie, Olifants, Letaba, Crocodile and Luvuvhu/Limpopo complex (Viljoen & Biggs, 1998). The total in the first three of these rivers averaged 2000 over the 11 year period with a range of 1675 to 2244. There were no significant differences between years except on Sabie River but the fluctuations there did not reveal any consistent trends. Numbers in Natal outside the Kruger National Park amount to about 1500. The population in Natal and Kwazulu (Zululand) has been monitored since 1951 (Taylor, 1987). Numbers have risen from a few hundred in the 1950s to 1264 in 1986 although much of the increase was due to the incorporation of new waters into the sample. Nevertheless, clear increases were seen in individual lakes such as Kobi Lake, Lake Lucia and Lake Sibaya as well as in the Ndumu Game Reserve. With adjustments for under-counting in the early days, Taylor considers that five-year averages have risen for the whole of Zululand from 664 in the late 1950s to 1423 for the period 1982 to 1986. Viljoen (1980) counted 305 and 34 hippos respectively in 1978 on stretches of the Olifants and Blyde Rivers to the east of the central Kruger National Park. These were not covered by Taylor's censuses. The sum of the estimates from various parts of the country exceeds 4500 and an overall total of 5000 is allowed for the whole of South Africa.

Conservation status: The conservation status of hippos is generally satisfactory in South Africa. Numbers fluctuate from year to year in the Kruger National Park but the population is more or less stable and is not giving cause for concern. The increases in Natal, however, are largely confined to the reserves and the Mkuzi Swamp system. Elsewhere there are declines, most notably in the Pongolo population. The species is fully protected and regulations are strictly enforced. Some culling has taken place in the perennial rivers in the Kruger National Park and, since 1977, in Ndumu Reserve and Lake St Lucia in Natal. The latter has averaged 70 a year with a range of 5–140, a rate which is unlikely to have had a significant effect on the survival of the species.

Zambia

Distribution: The hippo is widespread throughout Zambia on lakes and rivers including Lake Bangweulu, the Kafue River and Flats, Lake Kariba, Luangwa River, Lake Mweru, Lake Mweru Wantipa, Lake Tanganyika and the Zambezi River. A distribution map, given by Ansell (1978), shows that almost all rivers and lakes carry hippos.

Numbers; The populations in Zambia have been well surveyed, particularly those in the Luangwa Valley, and the results of counts in various regions have been published (Atwell, 1963; Ansell, 1965; Marshall & Sayer, 1976; Tembo, 1987; Norton, 1988). Marshall & Sayer (1976) surveyed the hippos along parts of the Luangwa River and showed that there had been an erratic increase between 1952 (337 hippos) and 1972 (1681 hippos). The species was very rare in the Luangwa Valley before the Second World War but whether this was due to a rinderpest outbreak or to excessive hunting is not clear. Norton (1988) carried out an aerial survey in 1981 along a 580 km stretch of the river from its confluence with the Lufila River to Luembe and returned a total of 14 250, which was a marked increase over previous counts on all sections of the river. Tembo (1987) made further aerial and ground counts in 1982 and 1983 although over a shorter length of river (165 km) than that covered by Norton and over different sections from those covered by Marshall & Sayer. His estimates for the two years were 6293 and 6544 respectively. Comparisons with previous unpublished reports on the files of the Zambian National Parks and Wildlife Service showed that the population had increased at a rate of 7% per annum since 1970. R.H.V. Bell (*in litt.*) gave an estimate of 20 000 to 25 000 as the population of the whole of the Luangwa Valley. This was based on the long series of ground counts in the South Luangwa National Park with an extrapolation to the rest of the Valley. He considers the Valley population to have increased at 15% p.a. since the late 1950s until there was a major die-off in 1987/88 due to anthrax, which caused the deaths of about 25% of the population. This reduction has now been more or less replaced by subsequent increases. The hippo is also very common in other regions of Zambia. Unpublished reports of aerial and ground surveys made in 1989 by R.C.V. Jeffery and his team on behalf of the National Parks Service record the widespread presence of hippos in the Kafue Flats, although the technique did not lend itself to accurate censussing. They also saw hippos in the Bangweulu Swamps and considered that there were many more than were observed. An estimate of 40 000 for the whole country is given by F.E.C. Munyenyembe (*in litt.*), with about half living in the Luangwa Valley. This figure is accepted here.

Conservation Status: With such a large population and a general increase in numbers, the conservation status of hippos in Zambia can only be satisfactory. There is some concern in the north-western provinces over excessive hunting according to F.E.C. Munyenyembe (*in litt.*). Hunting under licence is permitted outside protected areas but it does not seem to have made any serious inroads into hippo numbers over the country as a whole. Some authorities feel that further culling may be necessary in parts of the Luangwa Valley, given the 15% annual increase. There are plenty of hippos in protected areas although law enforcement is good only in the central regions of the Luangwa Valley. Conservation there is helped by the fact that most local people do not eat hippo because they fear that it causes leprosy. This may be a reflection of the more likely effects of eating hippos afflicted with anthrax. Elsewhere in Zambia, the hippo is eaten without inhibitions. The main threat to hippo results from the damage it causes to crops and fishing interests. The hippo has a considerable commercial value with its hide, teeth and meat and there is evidence of an increase in illegal hunting as a consequence (R.H.V. Bell, *in litt.*).

Zimbabwe

Distribution: Hippos occur on most large rivers in Zimbabwe, particularly the Zambezi, Limpopo and the Sabi/Lundi systems. They are common on Lake Kariba. They are also found in dams and on smaller rivers where there is permanent water. They may wander over large distances and be responsible for isolated records of the species. A map showing the distribution is included in Smithers & Wilson (1979).

Numbers: There has not been a country-wide survey of hippos but there are a number of local counts. An aerial survey in 1985 of about 1000 km of rivers in the southeast lowveld returned a total of 900 hippos (S.S. Towindo, *in litt.*). A count along a 50 km section of the Zambezi River produced 2000 (B. Child, *in litt.*). A 1990 survey of Chete Safari Area made by R. Johnson (*in litt.*) gave a minimum of 107 and a maximum of 128 hippos. Numbers in the Hwange National Park are very low with the Mandavu Dam being the only place where hippos are permanent residents (Wilson, 1975). No correspondent ventured a country-wide total but 7000 is probably not far off the mark.

Conservation status: The status of the hippo in Zimbabwe is generally satisfactory and in places it is considered to be over-populated. Some culling has been carried out in the lower Lundi river system and it is considered necessary in the Zambezi, where the high densities are believed to be responsible for increased mortality and to have adversely affected other grazing species (B. Child, *in litt.*). The hippos in the southeastern lowveld are thought to be threatened owing to conflicts with large irrigation schemes and some have been shot as problem animals. The hippo is fully protected in national parks but it may be shot in hunting areas although the quotas set are very low. Laws are generally enforced and any utilisation of hippos is well controlled. The tusks are used in carving. Some hippos have been introduced into numerous farm dams for hunting or as tourist attractions. There does not appear to be any serious threat to the survival of hippos in Zimbabwe apart from the risk of anthrax in the Zambezi river system.

SUMMARY OF THE CONSERVATION STATUS OF HIPPOS IN SOUTHERN AFRICA

The estimated number of hippos in each of the southern African countries is summarised in Table 12.1. The distribution is shown in Fig. 12.1. Of the eight countries considered, the hippo population is said to be decreasing in two, stable in three and increasing in one. In the two remaining countries, the trends are unknown. The situation does not appear to have led to many worries for only in two countries was concern expressed over the survival of the species. Some states hold very large numbers of hippos, the chief being Zambia with 40 000, Mozambique with 18 000 and Malawi with 10 000. The distribution is also widespread in those three countries. Legislation for the protection of the species is satisfactory and enforcement of the law is generally good or better.

Table 12.1   Estimates of hippo numbers in Southern Africa.

Country	Numbers	Conservation Status	Population Trend	Protection/ Enforcement	Concern?
Angola	2000	No recent information			
Botswana	4000	RD-LD	D	G/F	N
Malawi	10 000	W-LD	S	H/G	N
Mozambique	18 000	W-LA	D	H/U	Y
Namibia	400	RD-LA	U	G/G	Y
South Africa	5000	RD-LA	S	G/E	N
Zambia	40 000	W-LA	I	H/F	N
Zimbabwe	7000	RD-LA	S	H/E	N

Total for Southern Africa – 86 400

KEY
Conservation Status: LA = locally abundant; LD = low density; HD = high density; RD = restricted distribution; W = widespread.
Population Trend: D = decreasing; I = increasing; S = stable; U = unknown.
Protection: G = total protection; H = partial protection; U = unknown.
Enforcement: E = excellent; F = fair; G = good; P = poor; U = unknown.
Concern? (over population status): Y = yes; N = no.

CONCLUSIONS ON THE STATUS OF HIPPOS THROUGHOUT AFRICA

A summary of the counts for the whole of Africa is given in Table 12.2. The total of 173 690 is somewhat higher than the 157 000 that I estimated in an earlier analysis of the same data (Eltringham, 1993a) and no doubt a re-analysis would produce yet another figure, which would be as inaccurate as these two. The important point is the order of magnitude of the assessment. The techniques used did not allow for the degree of error to be determined but the true total is probably not far off these figures, i.e. of the order of a few hundred thousand. This might seem a comfortable total but

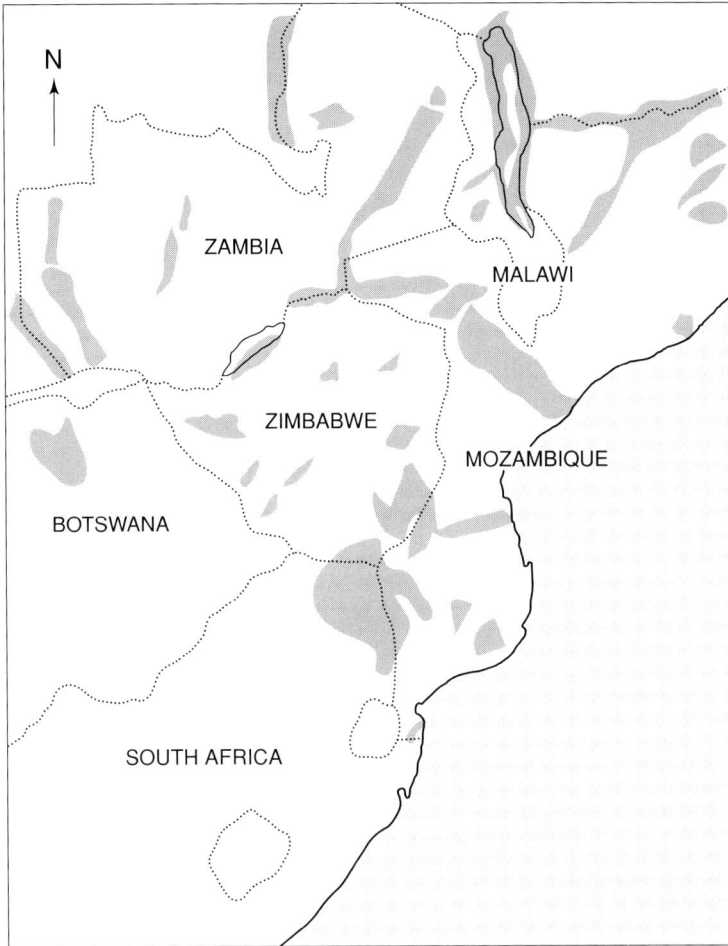

Figure 12.1 The distribution of the common hippopotamus in southern Africa as determined from the IUCN/SSC investigation.

Table 12.2 Estimate of hippo numbers in the whole of Africa.

Region	Number of Hippos
West	7740
East	79 550
Southern	86 400
Grand total	**173 690**

many of the populations are unlikely to be viable in the long-term and the future of the hippo needs more attention than it has received in the past. The future of the African elephant has attracted great publicity and it might appear from the public concern that the species is close to extinction. Yet the most pessimistic estimate of the number of elephants in Africa is half a million, more than twice our estimate of hippos. At the time my survey was carried out, the hippo had no official protection whatever outside the range states, some of which had passed laws for its preservation. The hippo did not appear on any of the appendices of CITES, the international agreement to control wildlife trade, except in Ghana, where it was listed on the ineffectual Appendix III. Following the publication of the report (Oliver, 1993), however, the hippo was included on Appendix II, which precludes trade in the animal or its products unless this can be shown not to be deleterious to the species' survival.

In 18 of the 34 countries for which data were obtained in the survey reported here, the hippo was thought to be declining in numbers, although in four of them some doubt was expressed over the reality of the decline.

Increases were reported in only two countries, Congo (Brazzaville) and Zambia. With only 50 hippos, Congo is not significant but Zambia's population is very much so as it is the country with the largest number of hippos. Populations were said to be stable in seven countries but no opinion was offered on possible trends in numbers for six of the states. No information was available for the remaining country, Rwanda, but in view of the recent warfare there, it may be stated with some confidence that numbers have declined.

This does not sound like an optimistic assessment of the future of the hippopotamus but the situation may not be as gloomy as it appears. Although, overall, numbers may be declining in a country, some, at least, of the reduced populations are maintaining themselves, e.g. the hippos in the Queen Elizabeth Park in Uganda. Other countries, such as Zaire, in which the decline is probably real, are at least starting from a high base-line and, although one shouldn't say so, they can afford to lose many animals before the situation becomes serious. Nevertheless, the level of attrition of these populations cannot go on indefinitely and the situation cannot be allowed to drift.

The greatest concern over the future of hippos is undoubtedly in West Africa. One should not expect too many there for much of the area is, or was, occupied by rain forest, which is not suitable habitat for hippos. The total number of hippos is indeed very small but probably it was once not quite as small as it is now. The low numbers are bad enough but worse still, those that remain are split into small, isolated groups of a few dozen at most, many of which are showing declines. On top of everything else, law enforcement is weakest in the West African states. Once populations have fallen to the extent that the hippos have in West Africa, genetic problems are likely to arise in the long term due to the increased risk of inbreeding. It might be thought that the West African hippos are not worth bothering about but their loss would mean the extinction of at least one subspecies (*H. a. tschadensis*). Even though the taxonomy may be flawed, these isolated hippos do form distinct genetic groups and their loss would deplete the species' gene pool.

The position is much brighter in East Africa, where there are some flourishing populations, such as those within Tanzania and Zaire. Zaire's future is problematic but Tanzania is a stable state with good conservation regulations that are reasonably well enforced. It is in such countries that most conservation effort should be directed. It is reassuring to note that in most of the countries, the hippos are locally abundant even where the total numbers are relatively low. This should mean that genetic problems are unlikely to arise.

The southern African countries are even more important than those in East Africa for the conservation of hippos. Malawi, Mozambique and Zambia have really large populations which are widespread and locally abundant. Protection and law enforcement could be better but are reasonably effective. In two of the countries where law enforcement is good, South Africa and Zimbabwe, the hippo populations are stable and are of a fair size.

One can conclude from this survey that, over the whole of Africa, there is no immediate threat to the hippo as a species although some of the constituent populations are certainly at risk.

REFERENCES

Abaturov, B.D., Kassaye, F., Kuznetsov, G.V., Magomedov, M-R.D. & Petelin, D.A. (1995) Nutritional estimate of populations of some wild free-ranging African ungulates in grassland (Nechisar National Park, Ethiopia) in dry season. *Ecography*, **18**:164–172.

Allbrook, D.B., Harthoorn, A.M., Luck, C.P. & Wright, P.G. (1958) Temperature regulation in the white rhinoceros. *J. Physiol.*, **143**:51–52P.

Anderson, S. & Jones, J.K. (1984) *Orders and Families of Recent Mammals of the World*. Wiley, New York.

Anon (1993) Hippos return to Sundays River. *Afr. Wildl.*, **47**:31.

Ansell, W.F.H. (1965) Hippo census on the Luangwa River, 1963–1964. *Puku*, **3**:15–27.

Ansell, W.F.H. (1978) *The Mammals of Zambia*. National Parks & Wildlife Service, Chilanga.

Anstey, S (1991) Large mammal distribution in Liberia. WWF-International, Gland.

Arman, P. & Field, C.R. (1973) Digestion in the hippopotamus. *E. Afr. Wildl. J.*, **11**:9–17.

Atwell, R.I.,G. (1963) Surveying Luangwa hippo. *Puku*, **1**:29–49.

Atwell, R.I.G. (1966) Oxpeckers and their associations with mammals in Zambia. *Puku*, **4**:17–48.

Baker, J.R. (1969) Trypanosomes of wild mammals in the neighbourhood of the Serengeti National Park. *Symp. Zool. Soc. Lond.*, **24**:147–158.

Barber, K.B., Buchanan, S.A. & Galbreath, P.F. (1980) *An Ecological Survey of the St. Floris National Park, Central African Republic*. International Park Affairs Division, National Park Service, Washington.

Barklow, W. (1995) Hippo Talk. *Nat. Hist.*, **5/95**:54.

Benedict, F.G. (1936) *The Physiology of the Elephant*. Carnegie Institute, Washington.

Bere, R.M. (1959) Queen Elizabeth National Park: Uganda. The hippopotamus problem and experiment. *Oryx*, **5**:116–124.

Bhima, R. (1996) Census of hippopotamus (*Hippopotamus amphibius* (L.)) in the Upper Shire River, Malawi. *Afr. J. Ecol.*, **34**:83–85.

Bishop, W.W. (1962) Gully erosion in the Queen Elizabeth National Park. *Uganda J.*, **26**: 161–165.

Black, J.C. & Sharkey, M.J. (1970) Reticular groove (*Sulcus reticuli*): An obligatory adaptation in ruminant-like herbivores. *Mammalia*, **34**:294–302.

Bourlière, F. & Verschuren, J. (1960) Introduction a l'ècologie des ongles du Parc National Albert. *Explor. Parc. nat. Albert*, Bruxelles.

Bousquet, B. & Szaniawski, A. (1981) Resultats des inventaires aeriens des grand mammiferes dans la region "Pendjari-Mekrou". Document de Terrain No. 4/ FO:DP/UPV/78/008. FAO. Ouagoadougou.

Carpaneto, G.M. & Germi, P. (1989) The mammals in the zoological culture of the Mbuti pygmies of north eastern Zaire. *Hystrix* (ns) **1**:1–83.

Chilvers, H.A. (1931) *Huberta Goes South, a Record of the Lone Trek of the Celebrated Zululand Hippopotamus, 1928–1931*. Gordon & Gotch, London.

Clemens, E.T. & Maloiy, G.M.O. (1982) The digestive physiology of three East African herbivores: the elephant, rhinoceros and hippopotamus. *J. Zool., Lond.*, **198**:141–156.

Clough, G. (1970) A record of "testis cords" in the ovary of a mature hippopotamus (*Hippopotamus amphibius* Linn). *Anatom. Rec.*, **166**:47–50.

Coe, M. & Collins, N.M. (1986) *Kora: An Ecological Inventory of the Kora National Reserve, Kenya*. Royal Geographical Society, London.

Coryndon, S.C. (1977) The taxonomy and nomenclature of the Hippopotamidae (Mammalia, Artiodactyla) and a description of two new fossil species. *Proc. Kon. Ned. Akad. Wetensch.* Ser. B., **80**:61–88.

Coryndon, S.C. (1978) Hippopotamidac. In: *Evolution of African Mammals* (Eds V.J. Maglio & H.B.S. Cooke), Harvard University Press, Cambridge. 483–485.

Cott, H.B. (1961) Scientific results of an enquiry into the ecology and economic status of the Nile Crocodile (*Crocodilus niloticus*) in Uganda and Northern Rhodesia. *Trans. zool. Soc., Lond.*, **29**:211–356.

Crisp, E. (1867) On some points connected with the anatomy of the hippopotamus *(Hippopotamus amphibius). Proc. zool. Soc. Lond.,* **39:**601–612.

Cristino, J.J. & Melo, de SA E. (1958) Statut des ongules en guinee Portugaise. *Mammalia,* **22:**387–389.

Curry-Lindahl, K. (1961) Contribution à l'ètude des vertèbrès terrestres en Afrique tropicale. *Inst. Parcs Nat. Congo Ruanda, Bruxelles.*

Dean, W.R.J. & MacDonald, I.A.W. (1981) Review of African birds feeding in association with mammals. *Ostrich,* **52:**135–155.

Despres, L., Kruger, F.J., Imbertestablet, D. & Adamson, M.L. (1995) ITS2 Ribosomal-RNA indicates *Schistosoma hippopotami* is a distinct species. *Int. J. Parasit.,* **25:**1509–1514.

Dinnik, J.A., Walker, J.B., Barnett, S.F. & Brocklesby, D.W. (1963) Some parasites obtained from game animals in western Uganda. *Bull. Epiz. Dis. Afr.,* **11:**37–44.

Dittrich, L. (1976) Age of sexual maturity in the hippopotamus *(Hippopotamus amphibius). Int. Zoo. Yb.,* **16:**171–173.

Dudley, J.P. (1996) Record of carnivory, scavenging and predation for *Hippopotamus amphibius* in Hwange National Park, Zimbabwe. *Mammalia,* **60:**486–490.

Duncan, W.R.H. & Garton, G.A. (1968) The fatty acid composition and intramolecular structure of triglycerides from adipose tissue of the hippopotamus and the African elephant. *Comp. Biochem. Physiol.,* **25:**319–325.

Dunham, K.M. (1990) Fruit production by *Acacia albida* trees in Zambezi riverine woodlands. *J. trop. Ecol.,* **6:**445–457.

Dupuy, A.R. (1971) Le Niokolo-Koba: Premier Grand Parc National de la Rèpublique de Sènègal. G.I.A., Dakar.

Eltringham, S.K. (1974) Changes in the large mammal community of Mweya Peninsula, Rwenzori National Park, Uganda, following removal of hippopotamus. *J. appl. Ecol.,* **11:**855–865.

Eltringham, S.K. (1980) A quantitative assessment of range usage by large African mammals with particular reference to the effects of elephants on trees. *Afr. J. Ecol.,* **18:**53–71.

Eltringham, S.K. (1984) *Wildlife Resources and Economic Development.* John Wiley & Sons, Chichester.

Eltringham, S.K. (1993a) The common hippopotamus *(Hippopotamus amphibius).* In: *Pigs, Peccaries and Hippos* (Ed. W.R.L. Oliver), IUCN, Gland. 43–55.

Eltringham, S.K. (1993b) The pygmy hippopotamus *(Hexaprotodon liberiensis).* In: *Pigs, Peccaries and Hippos* (Ed. W.R.L. Oliver), IUCN, Gland. 55–60.

Eltringham, S.K. (1993c) Review of priorities for conservation action and future research on hippopotamuses. In: *Pigs, Peccaries and Hippos* (Ed. W.R.L. Oliver), IUCN, Gland. 61–65.

Erwee, H. (1996) Tales from the bush. *BBC Wildlife,* May 1996:98.

Fagotto, F. (1985) Larger mammals of Somalia in 1984. *Environ. Conserv.,* **12:**260–264.

Faure, M. & Guerin, C. (1990) *Hippopotamus laloumena nov. sp.* the third Holocene *Hippopotamus* species of Madagascar. *C. r. Acad. Sci. Ser. II,* **310:**1299–1305.

Field, C.R. (1968a) Methods of studying the food habits of some wild ungulates in Uganda. *Proc. Nutr. Soc.,* **27:**172–177.

Field, C.R. (1968b) The food habits of some wild ungulates in relation to land use and management. *E. Afr. agric. For. J.,* **33:**159–162 (Special issue).

Field, C.R. (1968c) A comparative study of the food habits of some wild ungulates in the Queen Elizabeth National Park, Uganda preliminary report. *Symp. zool. Soc. Lond.,* **21:**135–151.

Field, C.R. (1970) A study of the feeding habits of the hippopotamus *(Hippopotamus amphibius* Linn.) in the Queen Elizabeth National Park, Uganda, with some management implications. *Zool. Afr.* **5:**71–86.

Field, C.R. (1972) The food habits of wild ungulates in Uganda by analyses of stomach contents. *E. Afr., Wildl. J.,* **10:**17–42.

Field, C.R. & Laws, R.M. (1970) The distribution of the larger herbivores in the Queen Elizabeth National Park, Uganda. *J. appl. Ecol.,* **7:**273–294.

Games, I. (1990) *A Survey of Hippopotamus in the Selous Game Reserve, Tanzania.* Unpublished Report to the Director of Wildlife, Tanzania.

Gansberger, K. & Forstenpointner, G. (1995) On the existence of a gall bladder in the hippopotamus. *Wien. Tierarz. Monatssch.,* **82:**157–158.

Garnham, P.C.C. (1960) Blood parasites of hippos in Uganda. *E. Afr. Med. J.,* **37:**495.

Gasgoyn, M., Currant, A.P. & Lord, T.C. (1981) Ipswichian fauna of Victoria Cave and the marine palaeo-climatic record. *Nature, Lond.,* **294:**652–654.

Gatesby, J., Hayashi, C., Cronin, M.A. & Arctander, P. (1996) Evidence from milk casein genes that cetaceans are close relatives of hippopotamid artiodactyls. *Mol. Biol. Evol.,* **13:**954–963.

Gaziri, A.W. (1987) *Hexaprotodon sahabiensis* (Artyodacyla, Mammalia): a new hippotamus from Libya. In: *Neogene Paleontology and Geology of Sahabi*. (Ed. D.N.T. Boaz *et al.*), A.R. Liss Inc., New York.

Goss, L.J. (1960) Breeding notes on the hippopotamus (*Hippopotamus amphibius*) and the giraffe (*Giraffa camelopardalis*) at Cleveland Zoo. *Int. Zoo. Yb.*, **2**:90.

Green, A.A. (1979) Density estimate of the larger mammals of Arly National Park, Upper Volta. *Mammalia*, **43**:59–70.

Green, A.A. (1997) Hippopotami of the Pendjari Ecosystem (Benin and Burkino Faso) during the 1970's. *Nigerian Field*, **62**:130–139.

Greenwood, P.J. (1980) Mating systems, philopatry and dispersal in birds and mammals. *Anim. Behav.*, 1140–1162.

Greenwood, P.J. (1984) Mating systems and the evolutionary consequences of dispersal. In: *The Ecology of Animal Movement*. (Eds I.R. Swingland & P.J. Greenwood), Clarendon Press, Oxford. 116–131.

Gregory, P.A. (1985) Common sandpiper feeding from hippopotamus injuries. *British Birds*, **78**:400.

Grubb, P. (1993) The Afrotropical Hippopotamuses *Hippopotamus* and *Hexaprotodon*: anatomy and description. In: *Pigs, Peccaries and Hippos*. (Ed. W.R.L. Oliver), IUCN, Gland. 41–43.

Grzimek, B. (1988) *Grzimeks Enzyklopadie: Saugetiere*. Band V. Kindler Verlag GmbH, Munchen.

Guggisberg, C.A.W. (1961) *Simba: the Life of the Lion*. Timmings, Cape Town.

Guilbride, P.D.L., Coyle, T.J., McAnulty, E.G., Barber, L. & Lomax, G.D. (1962) Some pathogenic agents found in hippopotamus in Uganda. *J. Comp. Path. Therap.*, **72**:137–141.

Gwynne, M.D. & Bell, R.H.V. (1968) Selection of vegetation components by grazing ungulates in the Serengeti National Park. *Nature, Lond.*, **220**:390–393.

Hanks, J. (1979) *A Struggle for Survival*. Country Life Books, Feltham.

Harris, J.M. (1991) Family Hippopotamidae. In: *Koobi Fora Research Project. Vol. 3. The Fossil Ungulates*. (ed. J.M. Harris), Clarendon Press, Oxford.

Hasegawa, M. & Adachi, J. (1996) Phylogenetic position of cetaceans relative to artiodactyls – reanalysis of mitochondrial and nuclear sequences. *Mol. Biol. Evol.*, **13**:710–713.

Hediger, H. (1951) Observations sur la psychlogie animale dans les parcs Nationaeux du Congo Belge. Inst. Parcs Nat. Congo Belge, Bruxelles.

Hentschel, K. (1990) *Untersuchung zu Status, Okologie und Erhaltung des Zwergflusspferdes* (Choeropus liberiensis) *in der Elfenbeinkoste*. Dr. re. nat. thesis, University of Braunschweig.

Heslop, I.R.P. (1945) The pygmy hippopotamus in Nigeria. *Field (Nigeria)*, **185**:629–630.

Hillman, J.C. (1982) *Wildlife Information Booklet*. Department of Wildlife Management, Ministry of Wildlife Conservation & Tourism, Southern Region, Sudan.

Hofmann, R. (1974) *The Ruminant Stomach*. East African Literature Bureau, Nairobi.

Houtekamer, J.L. & Sondaar, P.Y. (1979) Osteology of the fore limb of the Pleistocene dwarf hippopotamus from Cyprus with special reference to phylogeny and function. *Proc. Kon. Ned. Akad. Wetench*. Ser. B, **82**:411–448.

Hugget, A. St. G. & Widdas, W.F. (1951) The relationship between mammalian foetal weight and conception age. *J. Physiol.*, **114**:306–317.

Irwin, D.M. & Arnason, U. (1994) Cytochrome *b* gene of marine mammals: Phylogeny and evolution. *J. Mammal. Evol.*, **2**:37–55.

IUCN/UNEP (1987) *The IUCN Directory of Afrotropical Protected Areas*. IUCN, Gland & Cambridge.

Karstad, E.L. & Hudson, R.J. (1984) Census of the Mara River hippopotamus (*Hippopotamus amphibius*), southwest Kenya, 1980–1982. *Afr. J. Ecol.*, **22**:143–147.

Kayanja, F.I.B. (1989) The reproductive biology of the male hippopotamus. *Symp. zool. Soc. Lond.*, **61**:181–196.

Kenyi, J.M. (1979) Bone collection from Rwenzori National Park, Uganda. *Afr. J. Ecol.*, **17**:123–125.

Kingdon, J.S. (1979) East African Mammals: *An Atlas of Evolution in Africa. Vol. 3B (Large Mammals)*. Academic Press, London.

Kingdon, J. (1997) *The Kingdon Field Guide to African Mammals*. Academic Press, London.

Klingel, H. (1988) Grossflusspferde (Gattung *Hippopotamus*). In: *Grzimeks Enzyklopadie*. **5**:64–79.

Klingel, H. (1991) The social organisation and behaviour of *Hippopotamus amphibius*. In: *African Wildlife: Research and Management*. (Ed. F.I.B. Kayanja & E.L. Edroma) International Council of Scientific Unions, Paris. 73–75.

Kofron, C.P. (1993) Behavior of Nile crocodiles in a seasonal river in Zimbabwe. *Copeia*, No. **2**:463–469.

Lang, E.M. (1975) *Das Zwergflusspferd* Choeropsis liberiensis. A Ziemsen Verlag, Wittenberg.

Lang, E.M., Hentschel, M. v K. & Bulow, W. (1988) Zwergflusspferde (Gattung *Choeropsis*). In *Grzimeks Enzykopadie: Saugetiere*. Band V., Kindler Verlag, Munich. 62–64.

Langer, P. (1976) Functional anatomy of the stomach of *Hippopotamus amphibius* L. 1758. *S. Afr. J. Sci.*, **72:**12–16.

Laws, R.M. (1963) *The Nuffield Unit of Tropical Animal Ecology First Annual Report May 1962–April 1963.* Unpublished.

Laws, R.M. (1968a) Dentition and ageing of the hippopotamus. *E. Afr. Wildl. J.*, **6:**19–52.

Laws, R.M. (1968b) Interactions between elephant and hippopotamus populations and their environments. *E. Afr. agric. for. J.*, **23:**140–147.

Laws, R.M. & Clough, G. (1966) Observations on reproduction in the hippopotamus *Hippopotamus amphibius* Linn. *Symp. zool. Soc. Lond.*, **15:**117–140.

Laws, R.M., Parker, I.S.C. & Johnstone, R.C.B. (1975) *Elephants and their Habitats: the Ecology of Elephants in North Bunyoro, Uganda.* Clarendon Press, Oxford.

Ledger, H.P. (1968) Body composition as a basis for a comparative study of some East African mammals. *Symp. zool. Soc., Lond.*, **21:**289–310.

Lock, J.M. (1972) The effects of hippopotamus grazing on grasslands. *J. Ecol.*, **60:**445–467.

Luck, C.P. & Wright. P.G. (1959) The body temperature of the hippopotamus. *J. Physiol.*, **147:**53P.

Luck, C.P. & Wright, P.G. (1964) Aspects of the anatomy and physiology of the skin of the hippopotamus (*H. amphibius*). *Quart. J. exp. Physiol.*, **49:**1–14.

McCarthy, T.S., Ellery, W.N. & Bloem, A. (1998) Some observations on the geomorphological impact of hippopotamus (*Hippopotamus amphibius* L.) in the Okavango Delta, Botswana. *Afr. J. Ecol.*, **36:**44–56.

Macdonald, A.A. & Hartman, W. (1983) Comparative and functional morphology of the stomach in the adult and newborn pygmy hippopotamus (*Choeropsis liberiensis*). *J. Morphol.*, **177:**269–276.

Macdonald, D.W. (ed) (1984) *The Encyclopedia of Mammals.* George, Allen & Unwin, London.

MacKie, C.S. (1976) Feeding habits of the hippopotamus on the Lundi River, Rhodesia. *Arnoldia,* **7:**1–16.

Malpas, R.C. (1978) *The Ecology of the African Elephant in Rwenzori and Kabalega Falls National Parks.* Ph.D. thesis, University of Cambridge, U.K.

Marks, M. (1976) *Large Mammals and a Brave People: Subsistence Hunters in Zambia.* University of Washington Press, Seattle.

Marshall, P.J. (1985) A new method of censusing elephants and a hippo census in Yankari Game Reserve. *Nigerian Field*, **50:**5–11.

Marshall, P.J. & Sayer, J.A. (1976) Population ecology and response to cropping of a hippopotamus population in eastern Zambia. *J. appl. Ecol.*, **13:**391–403.

Mkanda, F.X. (1994) Conflicts between hippopotamus (*Hippopotamus amphibius* L.) and man in Malawi. *Afr. J. Ecol.*, **32:**75–79.

Mkanda, F.X. & Kumchedwa, B. (1997) Relationship between crop damage by hippopotamus (*Hippopotamus amphibius* L.) and farmer complaints in the Elephant Marsh. *J. Afr. Zool.*, **111:**27–38.

Modha, M.L. (1968) Basking behaviour of the Nile crocodile on Central island, Lake Rudolf. *E. Afr. Wildl. J.,* **6:**81–88.

Moir, R.J. (1965) The comparative physiology of ruminant-like animals. In: *Physiology of Digestion in the Ruminant.* (Ed. R.W. Dougherty *et al.*), Butterworth, London. 1–14.

Mugangu, T.E. & Hunter, M.L. (1992) Aquatic foraging by hippopotamus on Zaire: response to a food shortage. *Mammalia*, **56:**345–349.

Murray, MG. & Illius, A.W. (1996) Multispecies grazing in the Serengeti. In: *The Ecology and Management of Grazing* (Ed. J. Hodgson & A.W. Illius). CAB International.

Muwazi, R.T. & Kayanja, F.I.B. (1991) Reproduction in the male hippopotamus (*Hippopotamus amphibius*): the epididymis. In: *African Wildlife: Research and Management* (Ed. F.I.B. Kayanja & E.L. Edroma). International Council of Scientific Unions, Paris. 71–72.

Ndhlovu, D.E. & Balakrishnan, M. (1991) Large herbivores in Upper Lupande Game Management Area, Luangwa Valley, Zambia. *Afr. J. Ecol.*, **29:**93–104.

Ngog Nye, J. (1988) Contribution à l'étude de la structure de la population des hippopotames (*Hippopotamus amphibius* L.) au Parc National de la Bénoué (Cameroun). *Mammalia*, **52:**149–158.

Norton, P.M. (1988) Hippopotamus numbers in the Luangwa Valley, Zambia, in 1981. *Afr. J. Ecol.*, **26:**337–339.

Nowak, R.M. & Paradiso, J.L. (1983) *Walker's Mammals of the World.* John Hopkins Univ. Press, Baltimore.

O'Connor, T.G. & Campbell, B.M. (1986) Hippopotamus habitat relationships on the Lundi River, Gonarezhou National Park, Zimbabwe. *Afr. J. Ecol.*, **24:**7–26.

Ogen-Odoi, A.A. & Dilworth, T.G. (1987) Effects of burning and hippopotamus grazing on savanna hare habitat utilization. *Afr. J. Ecol.*, **25:**47–50.

Oliver, W.L.R. (Ed) (1993) *Pigs, Peccaries, and Hippos: Status Survey and Conservation Action Plan.* IUCN, Gland.

Olivier, R.C.D. (1975) Aspects of skin physiology in the Pigmy hippopotamus *Choeropsis liberiensis. J. Zool.*, **176**:211–213.

Olivier, R.C.D. & Laurie, W.A. (1974a) Habitat utilization by hippopotamus in the Mara River. *E. Afr. Wildl. J.*, **12**:249–271.

Olivier, R.C.D. & Laurie, W.A. (1974b) Birds associating with hippopotamus. *Auk*, **91**:169–170.

Owen-Smith, R.N. (1988) *Megaherbivores: The Influence of Very Large Body Size on Ecology.* Cambridge University Press, Cambridge.

Petrides, G.A. & Swank, W.G. (1965) Population densities and the range carrying capacity for the large mammals in Queen Elizabeth National Park, Uganda. *Zool. Afr.*, **1**:209–225.

Pienaar, U. de V., van Wyk P. & Fairall, N. (1966) An experimental cropping scheme of hippopotami in the Letaba River of the Kruger National Park. *Kodoe*, **9**:1–33.

Pitchford, R.J. & Visser, P.S. (1981) *Schistosoma*, Weinlandi, 1858 from *Hippopotamus amphibius* Linnaeus, 1758, in the Kruger National Park. *Onderstepoort J. Vet. Res.*, **48**:181–184.

Pitman, C.R.S. (1962) Albino hippopotamus. *Proc. zool. Soc., Lond.*, **139**:531–534.

Plowright, W., Laws, R.M. & Rampton, C.S. (1964) Serological evidence for the susceptibility of the hippopotamus (*Hippopotamus amphibius* Linnaeus) to natural infection with rinderpest virus. *J. Hyg., Camb.*, **62**:329–336.

Pooley, A.C. (1967) Bird/crocodile and bird/hippopotamus commensalism in Zululand. *Ostrich*, **38**:11–12.

Powell, C.B. (1993) *Sites and Species of Conservation Interest in the Central Axis of the Niger Delta.* Report of Recommendations to the National Resources Conservation Council, Nigeria.

Randi, E., Lucchini, V. & Diong, C.H. (1996) Evolutionary genetics of the Suiformes as reconstructed using mtDNA sequencing. *J. Mamm. Evol.*, **3**:163–194.

Ritchie, J. (1930) Distribution of the pygmy hippopotamus. *Nature, Lond.*, **126**:204–205.

Robinson, P.T. (1970) *The Status of the Pygmy Hippopotamus and other Wildlife in West Africa.* M.S. thesis, University of Michigan.

Robinson, P.T. (1981) The reported use of denning structures by the pygmy hippopotamus *Choeropsis liberiensis. Mammalia*, **45**:506–508.

Roth, H.H., Hoppe-Dominik, B., Muhlenberg, M. & Steinhauer-Burkart, B. (1996) *Repartition et Statut des Especes de Grands Mammiferes en Côte d'Ivoire. Partie V: Les hippopotames.* Zentrum fur Naturschutz, University of Gottingen.

Ruggiero, R.G. (1991) Prey selection of the lion (*Panthera leo* Linn.) in the Manovo-Gounda-St. Floris National Park, Central African Republic. *Mammalia*, **55**:23–33.

Ruggiero, F.G. (1996) Interspecific feeding associations: Mutualism and semi-parasitism between Hippopotami *Hippopotamus amphibius* and African Jacanas *Actophilornis africanus. Ibis*, **138**:346–348.

Sayer, J.A. & Green, A.A. (1984) The distribution and status of large mammals in Benin. *Mamm. Rev.*, **14**:37–50.

Sayer, J.A. & Rakha, A.M. (1974) The age of puberty of the hippopotamus (*Hippopotamus amphibius* Linn.) in the Luangwa River in eastern Zambia. *E. Afr. Wildl. J.*, **12**:227–232.

Schomburg, H. (1912) On the trail of the pygmy hippo. *Bull. N.Y. Zool. Soc.*, **16**:880–884.

Senior M. & Tong, E.H. (1963) Parturition in a hippopotamus. *Proc. zoo. Soc., Lond.*, **140**:57–59.

Sidney, J. (1965) The past and present distribution of some African ungulates. *Trans. zool. Soc., Lond.*, **30**:1–397.

Sikes, S.K. (1974) Wildlife conservation with reference to Benue Plateau State, Nigeria. *Nigerian Field*, **38**:67–70.

Simmons, A.H. (1998) Extinct pygmy hippopotamus and early man in Cyprus. *Nature, Lond.*, **333**:554–557.

Skinner, J.D. & Smithers, R.H.N. (1990) *The Mammals of the Southern African Subregion.* University of Pretoria, Pretoria.

Smithers, R.H.N. (1971) The mammals of Botswana. *Mus. mem. Nat. Mus. Mon. Rhodesia*, **4**:1–340.

Smithers, R.H.N. & Wilson, V.J. (1979) Check list and atlas of the mammals of Zimbabwe Rhodesia. *Mus. Mem. Nat. Mus. Mon. Rhodesia*, **9**:1–147.

Smuts, G.L. & Whyte, I.J. (1981) Relationships between reproduction and environment in the hippopotamus *Hippopotamus amphibius* in the Kruger National Park. *Koedoe.*, **24**:169–185.

Spinage, C.A. (1980) Parks and reserves in Congo Brazzaville. *Oryx*, **15**:292–295.

Spinage, C.A., Loevinsohn, M.E. & Ndoute, J. (1977) Etudes additionelles du Parc National Bamingui-Bangoran. Document de Travail No 8, CAF/72/010. FAO, Rome.

Spinage, C.A., Guinness, F., Eltringham, S.K. & Woodford, M.H. (1972) Estimation of large mammal numbers in the Akagera National Park and Mutara Hunting Reserve, Rwanda. *Terre et Vie*, **4**:561–570.

Stewart, D.R.M. (1965) The epidermal characters of grasses, with special reference to East African plains species. *Bot. Jb.*, **84**:63–116.

Stewart, D.R.M. & Stewart, J. (1963) The distribution of some large mammals in Kenya. *J. E. Afr. Nat. Hist. Soc.*, **24(3)**:1–52.

Stewart, D.R.M. & Talbot, L.M. (1962) Census of wildlife on the Serengeti, Mara and Loita Plains. *E. Afr. agric. for. J.*, **28**:56–60.

Stuart, A.J. (1986) Pleistocene occurrence of hippopotamus in Britain. *Quartarpalaontologie*, **6**:209–217.

Stuenes, S. (1989) Taxonomy, habits and relationships of the subfossil Madagascan hippopotami *Hippopotamus lemerlei* and *H. madagascariensis*. *J. Vert. Paleont.*, **9**:241–268.

Suzuki, K. (1997) *An Ecological Study of Luangwa's Hippos Based on Samples Culled in 1996*. Unpubl. report.

Suzuki, K. & Imae, H. (1996) *The Report on the Hippo Population Collected in Hippo Project 1995 in the Luangwa River*. Unpubl. report.

Taylor, R.H. (1987) *Monitoring of Hippopotamus Populations in Natal and KwaZulu*. Unpubl. report.

Tembo, A. (1987) Population status of the hippopotamus on the Luangwa River, Zambia. *Afr. J. Ecol.*, **25**:71–77.

Thornton, D.D. (1971) The effect of complete removal of hippopotamus on grassland in the Queen Elizabeth National Park, Uganda. *E. Afr. Wildl. J.*, **9**:47–55

Thurston, J.P. (1968a) The larva of *Oculotrema hippopotami* (Monogenea: Polystomatidae). *J. Zool., Lond.*, **154**:475–480.

Thurston, J.P. (1968b) The frequency distribution of *Oculotrema hippopotami* (Monogenea: Polystomatidae). *J. Zool., Lond.*, **154**:481–485.

Thurston, J.P., Noirot-Timothee, C. & Arman, P. (1968) Fermentative digestion in the stomach of *Hippopotamus amphibius* (Artiodactyla : Suiformes) and associated ciliated protozoa. *Nature, Lond.*, **218**:882–883.

Tobler, K. (1988) *International Studbook for the Pygmy Hippopotamus* Choeropsis Liberiensis (Marten 1844). The Zoological Garden, Basel Zoo, Basel.

Turnbull, P.C.B., Bell, R.H.V., Sargawa, K., Munyenyembe, F.E.C., Mulenga, C.K. & Makala, L.H.C. (1991) Anthrax in wildlife in the Luangwa Valley, Zambia. *Vet. Rec.*, **128**:399–403.

Van Hoven, W. (1978) Digestion physiology in the stomach complex and hind gut of the hippopotamus (*Hippopotamus amphibius*). *S. Afr. J. Wildl. Res.*, **8**:59–64.

Verheyen, R. (1954) Monographie ethologique de l'hippopotame (*Hippopotamus amphibius* Linn.). *Explor. Parc. nat. Albert, Brussels*, 1–91.

Verschuren, J, Heymans, J.C. & Delvingt, W. (1989) Conservation in Benin – with the help of the European Economic Community. *Oryx*, **23**:22–26.

Viljoen, P.C. (1980) Distribution and numbers of the hippopotamus in the Olifants and Blyde Rivers. *S. Afr. J. Wildl. Res.*, **10**:129–132.

Viljoen, P.C. & Biggs, H.C. (1998) Population trends of hippopotami in the rivers of the Kruger National Park, South Africa. In: *Behaviour and Ecology of Riparian Mammals*. (Ed. N. Dunstone & M. Gorman). Cambridge University Press, Cambridge.

Weiler, P., De Meulenaer, T. & Vanden Bloock, A. (1994) Recent trends in international trade in hippopotamus ivory. *TRAFFIC Bull.*, **15**:47–49.

Wilson, V.J. (1975) Mammals of the Wankie National Park, Rhodesia. *Mus. mem. Nat. Mus. Monum. Rhodesia.* **5**:1–147.

Wright, P.G. (1964) Wild animals in the tropics. *Symp. zool. Soc. Lond.*, **13**:17–28.

Wright, P.G. (1987) Thermoregulation in the hippopotamus on land. *S. Afr. J. Zool.*, **22**:237–242.

Wyatt, J.R. (1971) Osteophagia in Masai giraffe. *E. Afr. Wildl. J.*, **9**:157.

Yalden, D.W., Largen, M.J. & Kock, D. (1984) Catalogue of the mammals of Ethiopia 5. Artiodactyla. *Monitore zoologico italiano*. Suppl. 25, No **4**:67–221.

Yoaciel, S.M. (1981) Changes in the populations of large herbivores and in the vegetation community in Mweya Peninsula, Rwenzori National Park, Uganda. *Afr. J. Ecol.*, **19**:303–312.

INDEX